Sampling Procedures to Detect Mycotoxins in Agricultural Commodities

Thomas B. Whitaker • Andrew B. Slate
M. Bruno Doko • Britt M. Maestroni
Andrew Cannavan

Sampling Procedures to Detect Mycotoxins in Agricultural Commodities

Springer

Thomas B. Whitaker
US Department of Agriculture
Agricultural Research Service
North Carolina State University
Raleigh, NC 27695-7625
USA
thomas_whitaker@ncsu.edu

Andrew B. Slate
Biological and Agricultural Engineering
Department
North Carolina State University
Raleigh, NC 27695-7625
USA
andrew_slate@ncsu.edu

M. Bruno Doko
Food and Environmental Protection
Laboratory
Joint FAO/IAEA Division of Nuclear
Techniques in Food and Agriculture
International Atomic Energy Agency
(IAEA)
Wagramer Strasse 5
P.O. Box 100, A1400 Vienna
Austria
Bruno.Doko@chello.at

Britt M. Maestroni
Food and Environmental Protection
Laboratory
Joint FAO/IAEA Division of Nuclear
Techniques in Food and Agriculture
International Atomic Energy Agency
(IAEA)
Wagramer Strasse 5
P.O. Box 100, A1400 Vienna
Austria
B.Maestroni@iaea.org

Andrew Cannavan
Food and Environmental Protection
Laboratory
Joint FAO/IAEA Division of Nuclear
Techniques in Food and Agriculture
International Atomic Energy Agency
(IAEA)
Wagramer Strasse 5
P.O. Box 100, A1400 Vienna
Austria
A.Cannavan@iaea.org

ISBN 978-90-481-9633-3 e-ISBN 978-90-481-9634-0
DOI 10.1007/978-90-481-9634-0
Springer Dordrecht Heidelberg London New York

Library of Congress Control Number: 2010936360

Cover illustration: "Milling a maize sample for fumonisin analysis" taken in the Food and Environmental Protection Laboratory, IAEA, Vienna, Austria

Printed on acid-free paper

Springer is part of Springer Science+Business Media (www.springer.com)

Preamble

The occurrence of mycotoxins in agricultural commodities has significant implications for the health of the consumer. Because of this, many countries have introduced regulatory limits for various mycotoxins in a range of products. Adherence to these limits is important to safeguard the consumer and also to permit trade in the affected commodities across international borders. Effective schemes to test for the presence and concentration of mycotoxins depend not only upon sound analytical methods, but also on sampling plans designed to ensure that the results of analysis of laboratory samples reflect, as accurately as possible, the overall concentrations in consignments or lots of produce, and to provide a good estimate of the variability or uncertainty associated with the analytical results. The results can then be used to implement regulatory decisions on the suitability of lots of produce for consumption or trade.

The purpose of this manual is to provide background information to food analysts and regulatory officials on effective sampling plans to detect mycotoxins in food. The reader is exposed to the concepts of uncertainty and variability in the mycotoxin test procedure as well as the importance of ensuring that samples are representative of the lot being sampled, and the consequences of a poorly designed sampling plan on the reliability of the measured levels of mycotoxins, possibly resulting in legal disputes and barriers to trade.

Currently, sampling guidelines and regulations are not harmonized throughout the world and there are various schemes in common use. This manual provides guidance on one approach. Practical guidance is also available in the training video presentation 'Sampling procedures for mycotoxins determination in food and feed products, including sampling and sample preparation', produced by the Food and Agriculture Organization of the United Nations (FAO) in conjunction with the Italian National Institute of Health.[1] This manual should be read in conjunction

[1] Sampling procedures for mycotoxins determination in food and feed products, including sampling and sample preparation (video presentation) http://www.soluzionepa.it/produzioneaudiovisivi.html

with the relevant legislation/guidelines on sampling for mycotoxins determination, including the General Guidelines on Sampling[2] and sampling plan for total aflatoxins in peanuts,[3] both published by Codex Alimentarius. Methods of sampling and analysis for the official control of the levels of mycotoxins in foodstuffs have also been described by the European Union.[4]

The lack of harmonization in the design of sampling plans is also reflected in the terminology used in the various texts. The terminology used in this manual is as defined by Codex Alimentarius in the sampling plan for total aflatoxins in peanuts.[3]

The views expressed in this text do not necessarily reflect those of the IAEA or FAO, or the governments of their Member States.

The mention of names of specific companies or products (whether or not indicated as registered) does not imply any intention to infringe proprietary rights, nor should it be construed as an endorsement or recommendation on the part of IAEA or FAO.

---

[2] Food and Agriculture Organization (2004). CAC/GL 50-2004, General Guidelines on Sampling, Viale della Terme di Caracalla, 00100, Rome, Italy, http://www.codexalimentarius.net/download/standards/10141/CXG_050e.pdf

[3] Food and Agriculture Organization (2001). CODEX STAN 209-1999, Rev. 1-2001, Maximum level and sampling plan for total Aflatoxins in peanuts intended for further processing, Viale della Terme di Caracalla, 00100, Rome, Italy, http://www.ipfsaph.org/cds_upload/kopool_data/codex_0/en_cxs_209e.pdf

[4] Commission Regulation (EC) No 401/2006 of 23 February 2006 laying down the methods of sampling and analysis for the official control of the levels of mycotoxins in foodstuffs, Official Journal of the European Union, L70/12, 9.3.2006, http://eur-lex.europa.eu/LexUriServ/LexUriServ.do?uri=OJ:L:2006:070:0012:0034:EN:PDF

Contents

Chapter 1
Introduction

Mycotoxins are chemically and biologically active secondary metabolites from several families of saprophytic and plant pathogenic moulds such as *Fusarium*, *Aspergillus*, *Penicillium*, *Alternaria*, and *Claviceps* spp. that grow on cereals, nuts, beans and many other agricultural crops including many fruit crops (Cullen and Newberne 1994). Mycotoxin residues in animal tissues, e.g. kidneys, and animal products, e.g. eggs, milk and dairy products can occur after ingestion of contaminated feeds by animals. Mycotoxin contamination of agricultural commodities occurs as a result of environmental conditions in the field as well as improper harvesting, storage, and processing operations. Crops grown and handled under warm and moist weather of tropical and subtropical countries are more prone to mycotoxin contamination than those in temperate regions, although certain fungi (e.g., *Fusarium* spp.) do grow at low temperature and produce toxins (e.g., trichothecenes).

Sound mycotoxin management should begin in the field before harvest, where the toxigenic fungi first become associated with the crop and where the contamination process begins. Pre-harvest invasion of agricultural crops by fungi is governed primarily by plant host–fungus and other biological interactions (e.g., insects). Post-harvest fungal growth is governed by the crop (nutrients) and by physical (temperature, moisture) and biotic (insects, interference competition) factors. Field fungi such as *Fusarium* and *Alternaria* species require high relative humidity and water content and are not competitive in storage systems (Moss 1991), which become dominated by storage fungi, particularly *Aspergillus* and *Penicillium* species that require minimal free water (Fig. 1.1). The mycotoxins most frequently encountered in cereals are listed in Table 1.1, and include, Fusarium toxins, i.e. fumonisins, trichothecenes (mainly deoxynivalenol and/or nivalenol) and zearalenone, and the Aspergillus and Penicillium toxins such as aflatoxins and ochratoxin A (Council for Agricultural Science and Technology 2003).

The *Fusarium* species are now recognized to be a major agricultural problem. They occur worldwide on a variety of plant hosts, primarily the cereal grains, producing a number of secondary metabolites of varied concentrations with widely divergent biological and toxicological effects in humans and animals after ingestion of fungus-damaged commodities. Mycotoxin entry to the human and animal dietary systems is mainly by ingestion but increasing evidence also points at entry by inhalation.

T.B. Whitaker et al., *Sampling Procedures to Detect Mycotoxins in Agricultural Commodities*,
DOI 10.1007/978-90-481-9634-0_1,

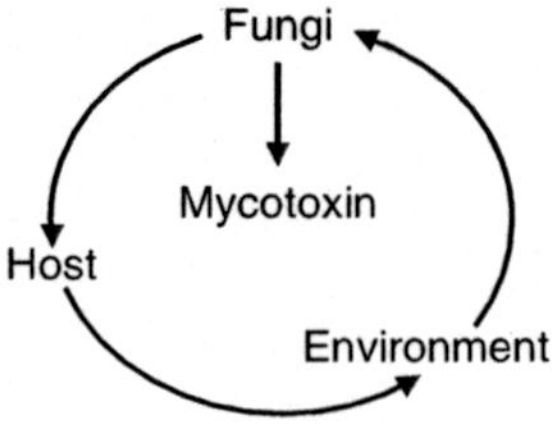

Fig. 1.1 The system of fungi, host, and environmental conditions must all be functioning to produce a mycotoxin

Table 1.1 The mycotoxins frequently encountered in cereals

The Mycotoxins	Major Mycotoxin Producers	Agricultural Commodity Frequently Contaminated
1. The *Fusarium* toxins		
Fumonisins[a]	*Fusarium moniliforme*, *F. proliferatum*	Maize
DON, and NIV[b,c]	*F. graminearum, F. culmorum,*	Maize, Wheat, Barley, Sorghum, Sunflower
	F. sporotrichioides, *F. crookwellence*	
Zearalenone (ZEA)	*F. graminearum, F. culmorum*	Maize, Oats, Barley, Malt, Sorghum, Rice
2. The *Aspergillus* and *Penicillium* toxins		
Aflatoxins[d]	*A. flavus,* A. *parasiticus*	Maize, Rice, Peanuts, Nuts, Spices, Milk (M_1)[e]
Ochratoxin A[f]	*A. ochraceus, Penicillium verrucosum*	Maize, Coffee, Wheat, Barley, Oats, Rye, Sorghum, Peanuts

[a] Of the six fumonisin analogues that are known at present, i.e., fumonisin A_1, (FA_1), A_2 (FA_2), B_1 (FB_1), B_2 (FB_2), B_3 (FB_3), and B_4 (FB_4), FB_1, FB_2 and FB_3 have been shown to occur naturally in maize, maize-based human foodstuffs and animal feeds worldwide

[b] Deoxynivalenol (DON) and Nivalenol (NIV) are the major natural trichothecenes contaminants in grains. *F. graminearum* was identified as the predominant DON and NIV producer. Another *Fusarium* species that produces DON is *F. culmorum*, whereas *F. sporotrichioides*, *F. crookwellence* and *F. poae* are found to be also NIV producers

[c] *F. graminearum* and *F. culmorum*, the most common pathogenic species are associated with cereals grown in warmer areas and in cooler areas, respectively

[d] Aflatoxins consist of a group of 20 related fungal metabolites, although only aflatoxins B_1, B_2, G_1 and G_2 are normally the major substances found in foods. The four major aflatoxins, i.e., B_1, B_2, G_1, G_2 (B = blue; and G = green fluorescence while the subscript designates relative chromatographic mobility), plus additional aflatoxin metabolic products, M_1 and M_2 are the most significant toxins as direct contaminants of foods and feeds

[e] The occurrence of M_1 in milk results from the ingestion of aflatoxin B_1-contaminated grains

[f] Of ochratoxins, the ochratoxin A (OTA), the 7-carboxy-5-chloro-8-hydroxy-3,4-dihy-dro-3-meth-yliso-coumarinamide of L-phenyalalanine, is the most toxic and is produced with the highest yield

The development of mycotoxins in stored products can be avoided by preventing the growth of toxin-producing molds. Mold growth can be prevented by ensuring that moisture and temperature conditions favorable to growth and proliferation do not occur.

Thus, as for the prevention of post-harvest mold infection and subsequent mycotoxin contamination, the control of physical (temperature, moisture) and biotic (insects, interference competition) factors remains of great importance (Phillips et al. 1994). In addition, in storage, the initial grain condition is critical as far as the fate of kernels is concerned. Good quality, clean, sound grain is easier to maintain in storage than physically damaged grain. Physically damaged kernels (i.e., insect damage and broken kernels) have a higher correlation with mold infestation and aflatoxin contamination than sound whole kernels.

Mycotoxin contamination of agricultural commodities and the subsequent impact on consumers' (humans and animals) health as well as on national and international trade are increasingly recognized in both developed and developing countries. Developed countries enact regulatory limits to protect consumers from exposure to mycotoxins. In many developing countries, however, regulation is insufficient and several agricultural commodities, including dietary staple foods, can contain unacceptably high levels of mycotoxins.

1.1 Regulatory Limits

Because mycotoxins are toxic and carcinogenic in animals, many countries regulate the maximum level that can occur in foods and feeds. Most regulations are concerned with controlling aflatoxin because it is considered the most toxic and carcinogenic of the naturally occurring mycotoxins. A recent FAO/WHO survey indicated that almost 100 countries regulate aflatoxin and several other mycotoxins in foods and feeds (Food and Agriculture Organization 2003). However, maximum levels differ widely from country to country because of a lack of agreement on what constitutes a safe maximum level for humans. Some of the maximum limits found in the FAO/WHO survey for aflatoxin are shown below in Table 1.2.

Because of differences in regulatory limits for mycotoxins, FAO and WHO, working through the CODEX system, are attempting to harmonize international maximum limits and sampling plans for mycotoxins to promote world trade and protect the consumer. CODEX has been successful in establishing a maximum limit for aflatoxin in raw shelled peanuts destined for further processing of 15 total ng/g and a sampling plan that uses a single 20 kg sample (Food and Agriculture Organization 2001a).

1.2 Mycotoxin Units

Mycotoxins are measured in concentration units or a ratio of the mass of the mycotoxin to the mass of the commodity. Units of measurement are usually grams of mycotoxin divided by the grams of commodity. Because the mass of a mycotoxin is usually very small, the units are reported in parts per billion (ppb) or parts per million (ppm). One ppb is 1 nanogram (ng) of mycotoxin per 1 g of commodity or

Table 1.2 **Examples of aflatoxin legal limits found in** various countries

	Aflatoxin Legal Limit[a] (ng/g)	
Countries	AF B_1	Total Aflatoxins
Africa		20
Egypt (for maize)		20
Egypt (for peanuts)		10
Nigeria	20	
South Africa		10
Australia		15
Canada		15
EU[b]	8	15
Philippines		20
US		20

[a]AF B_1 = Aflatoxin B_1 & Total Aflatoxins = AF (B_1 + B_2 + G_1 + G_2)
[b]Limits are for raw peanuts destined for further processing

Fig. 1.2 Several examples of the units equivalent to 1 ppb

- 1 ft in distance from earth to moon
- 1 cm in 10,000 km
- 1 second in 31.7 years

0.00000001 g/1 g or 1 ng/g. Several analogous examples of the magnitude of 1 ppb are illustrated in Fig. 1.2.

1.3 The Rationale of Accurate Sampling Procedures

Mycotoxins are unevenly distributed in grain storage bins or in the field, so that high concentrations of toxins could be found in "hot spots" or "pockets" in bulk storage of commodities or sometimes in a single ear or kernel of corn. Recent advances in methodology have been applied to mycotoxin analysis to significantly improve the capabilities for efficient detection and quantification of mycotoxins in agricultural commodities. In contrast, the development of sampling procedures remains a typical problem, although Codex and other bodies do have recommended sampling regimes.

In general, the suspected contaminated commodity should first be submitted to a rigorous sampling program in order to be diverted, if necessary, to protect the health of consumers. Inadequate sampling often yields meaningless data regardless of how good the approach is in the subsequent analysis. The collection of truly representative samples requires carefully designed sampling protocols. In regulatory operations, it is important to be able to measure accurately the true levels of a mycotoxin in a commodity so that correct decisions can be made about the fate of the product. However, an accurate estimation of the mycotoxin concentration in a large quantity of grain is difficult, owing to the large variability (or errors) associated with the mycotoxin test procedure, which includes sampling, sample preparation, and analytical steps. Because of the errors associated with each step of the testing procedure, the mycotoxin concentration in the lot[1] cannot be measured with absolute certainty and the result should always be reported with an estimate of the uncertainty.

A number of papers (Council for Agricultural Science and Technology 2003; Food and Agriculture Organization 2003) review regulatory limits and sampling of commodities for mycotoxin analysis used by various countries. For mycotoxins other than aflatoxins, general principles, schemes, and sampling plans adopted so far are those used for sampling plans and analysis of aflatoxins.

More work is needed in this area, especially in the refinement of sampling plans according to the real variability of certain mycotoxins in the food commodities.

[1]Lot: an identifiable quantity of a food commodity delivered at one time and determined by the official to have common characteristics, such as origin, variety, type of packing, packer, consignor or markings (Food and Agriculture Organization 2001b).

Chapter 2
Definition of a Sampling Plan

It is important to be able to detect and quantify the mycotoxin concentration in foods and feeds destined for human and animal consumption. In research, quality assurance, and regulatory activities, correct decisions concerning the fate of commercial lots can only be made if the mycotoxin concentration in the lot can be estimated with a high degree of accuracy and precision. The mycotoxin concentration of a lot is usually estimated by measuring the mycotoxin concentration in a small representative sample taken from the lot, called the laboratory sample[1] (Fig. 2.1).

Then, based on the measured laboratory sample concentration, a decision is made about the quality of the lot. For example, in a regulatory environment, decisions will be made to classify the lot as acceptable or unacceptable based upon a comparison of the measured sample concentration to a legal limit (the term "sample" by itself in this manual refers to a laboratory sample). If the sample concentration does not accurately reflect the lot concentration, then the lot may be misclassified and there may be undesirable economic and/or health consequences. Fortunately, sampling plans can be designed to minimize the misclassification of lots and reduce the undesirable consequences associated with regulatory decisions about the fate of bulk lots. In this manual, sampling plans will be defined, sources of uncertainty associated with a mycotoxin sampling plan will be identified, risks associated with misclassifying lots will be discussed, and methods that reduce misclassification of lots will be described.

A mycotoxin-sampling plan is defined by a mycotoxin test procedure and a defined accept/reject limit. A mycotoxin-test procedure is a multi-stage process (Fig. 2.2) and generally consists of three steps: sampling, sample preparation, and analysis (quantification).

---

[1]Laboratory sample: smallest size sample comminuted in a grinder (Food and Agriculture Organization 2001b).

T.B. Whitaker et al., *Sampling Procedures to Detect Mycotoxins in Agricultural Commodities*,
DOI 10.1007/978-90-481-9634-0_2,

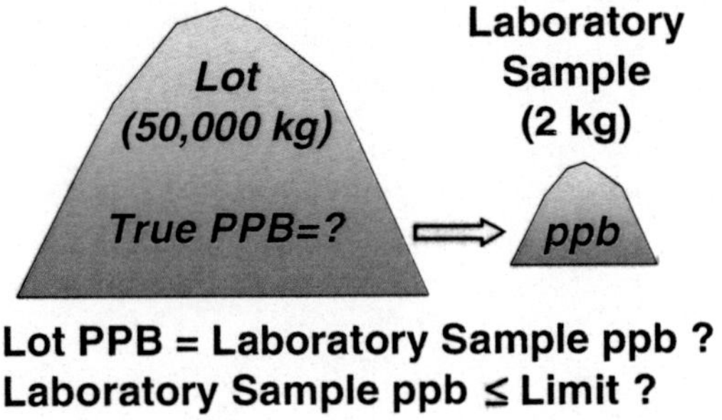

Fig. 2.1 Lot mycotoxin concentration is assumed to equal the measured mycotoxin concentration in a representative laboratory sample

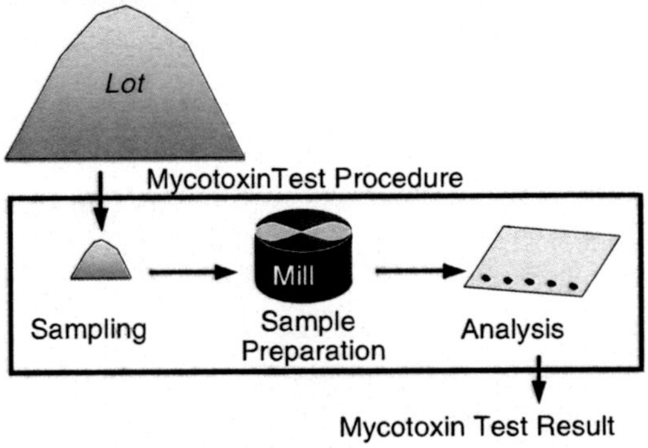

Fig. 2.2 A mycotoxin-test procedure usually consists of a sampling, sample preparation and analytical steps

The sampling step addresses issues such as when, how, how many. It specifies how the sample will be selected or taken from the bulk lot, the number of samples, and the size of the sample(s). For granular products, sample preparation includes the processing of the laboratory sample (i.e. grinding in a mill to reduce particle size) and the selection of a test portion, which is removed for subsequent analysis. Finally, in the analytical step, the mycotoxin is solvent extracted from the test portion and quantified using validated analytical procedures.

The measured mycotoxin concentration in the test portion is used to estimate the true mycotoxin concentration in the bulk lot or compared to a defined accept/reject limit that is usually equal to a maximum limit or regulatory limit. It is, therefore, important that the sampling procedure defines a laboratory sample that is as representative as possible of the bulk lot. Comparing the measured concentration in a test portion taken from a laboratory sample to an accept/reject limit is often called acceptance sampling because the actual measured concentration is not as important as whether that concentration, and thus the lot concentration, is above or below a legal limit. In activities other than regulatory acceptance sampling, for example in quality assurance or research, a precise and accurate estimate of the true lot mycotoxin concentration may be required.

According to Codex STAN209-1999, Rev. 1-2001 on "Maximum level and sampling plan for total Aflatoxins in peanuts intended for further processing" (Food and Agriculture Organization 2001b) and the Commission regulation (EC) 401/2006, on "Methods of sampling and analysis for the official control of the levels of mycotoxins in foodstuff" (Commision regulation (EC) 401/2006) each lot of materials which is to be examined must be sampled separately. Large lots should be subdivided into sublots[2] to be sampled separately. The subdivision can be done following the provisions laid down in Table 2.1.

Table 2.1 Subdivision of large lots into sublots for sampling

Lot Weight Tonne (T)	Weight or Number of Sublots	Number of Incremental Samples	Aggregate Sample Weight (kg)	Laboratory Sample Weight (kg)	Weight of Incremental Samples (g)
Peanuts (CODEX Standard)					
≥500	100 tonnes	100	20	20	
>100 and <500	5 sublots	100	20	20	
>25 and ≤100	25 tonnes	100	20	20	200
>15 and ≤25		100	20	20	
<15		10–100	≤20	20	
Groundnuts, pistachios, brazil nuts and other nuts (EC Regulation)					
≥500	100 tonnes	100	30	3 × 10 kg[a]	
>125 and <500	5 sublots	100	30	3 × 10 kg	
≥15 and ≤125	25 tonnes	100	30	3 × 10 kg	300
<15		10–100	≤30	3 × 10 kg	

[a]The division into three laboratory samples is not necessary in case of groundnuts and nuts subjected to further sorting or other physical treatment and of the availability of equipment which is able to homogenise a 30 kg sample.

[2]Sublot: designated part of a large lot in order to apply the sampling method on that designated part. Each sublot must be physically separate and identifiable.

Chapter 3
Uncertainty

There is always some level of uncertainty (variability) associated with any sampling plan. Because of this, the true mycotoxin concentration of a bulk lot can't be determined with 100% certainty; nor can all lots be correctly classified into good and bad categories (based upon some legal limit) with 100% accuracy. Accuracy and precision are two types of uncertainties associated with a sampling plan (Cochran and Cox 1957).

3.1 Accuracy

Accuracy is defined as the closeness of agreement between a measured value and an accepted reference value (ideally the true value). Another term associated with accuracy is "bias", which is the difference between the expectation of the test result (average m) and the accepted reference value. Using target practice as an example, the center of the target is analogous to the true value and holes in the target represent the measured values (Fig. 3.1).

Figure 3.1 shows that the rifle used on the left is not as accurate as the rifle used on the right where the average of the cluster of shots is around the center of the target. As shown in Eq. 3.1, accuracy (A) is the absolute difference between the true value (U) and the average of the n measured values (X_i)

$$\mathbf{A} = \left|\mathbf{U} - \left[\Sigma(\mathbf{X}_i)/\mathbf{n}\right]\right| \mathbf{for\ i = 1, 2, \ldots, n} \tag{3.1}$$

Biases have the potential to occur in the sample selection process, sample preparation process, and in the quantification steps of the mycotoxin test procedure. Biases should be the easiest component of uncertainty to control and reduce to acceptable levels, but methods to reduce bias are difficult to evaluate because of the difficulty in knowing the true mycotoxin concentration of the lot. Sample selection and sample preparation equipment and analytical methods must be continuously performance tested to minimize any biases. Before their application, analytical methods must undergo a validation process to show that they are 'fit for purpose'. This is ideally

T.B. Whitaker et al., *Sampling Procedures to Detect Mycotoxins in Agricultural Commodities*,
DOI 10.1007/978-90-481-9634-0_3, © Springer Science+Business Media B.V. 2010

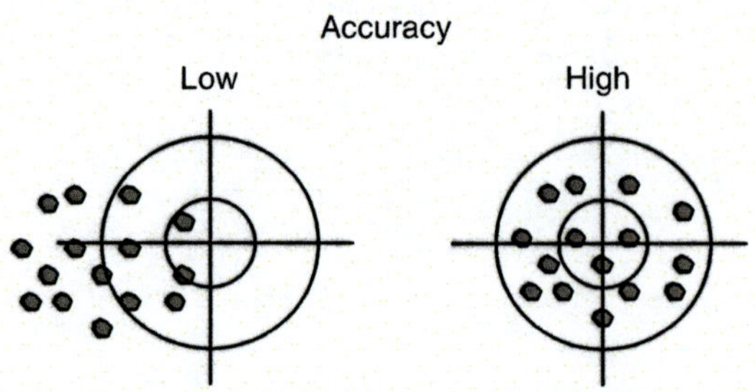

Fig. 3.1 Examples of low and high accuracy using target practice as an example

performed using certified reference materials containing a known concentration of the target substance(s), but can also be achieved by spiking samples of the commodity to be tested with a known concentration of the target substance. The validation process produces a set of parameters, which describes the performance of the analytical method in terms of its accuracy and precision (described below) and identifies and quantifies any bias that is inherent in the method. The performance of the method must be verified continuously using control samples in the laboratory, and through participation in proficiency tests or inter-laboratory comparisons.

3.2 Precision

Precision (variability) is defined as the closeness of agreement between independent test results obtained under stipulated conditions. The definition of precision makes no mention about how close the measured values are to the true value. Using target practice to illustrate precision, the closeness of the holes to each other is a measure of precision (Fig. 3.2).

Three statistical measures, variance (V), standard deviation (SD), or coefficient of variation (CV) can be used as a measure of precision (P).

$$\mathbf{V} = \left[\sum (\mathbf{x}_i - \mathbf{m})^2 / (\mathbf{n} - 1)\right] \textbf{ for i} = 1, 2, \ldots, \mathbf{n} \tag{3.2}$$

$$\mathbf{SD} = \sqrt{(\mathbf{V})} \tag{3.3}$$

The CV, expressed as a percent, is calculated as

$$\mathbf{CV} = \mathbf{100} * (\mathbf{SD}/\mathbf{m}) \tag{3.4}$$

Where x_i is the measured value and m is the mean of the x_i values.

Precision is a measure of variability. Variability can occur with each step of the mycotoxin test procedure and is usually associated with the mycotoxin distribution among contaminated particles in the lot. As will be shown later, increasing the quantity

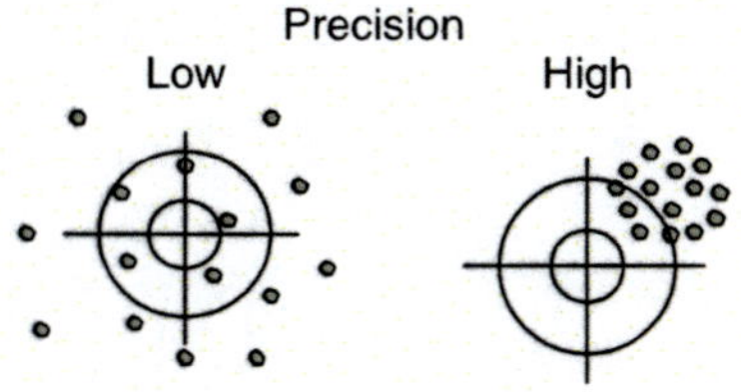

Fig. 3.2 Examples of low and high precision using target practice as an example

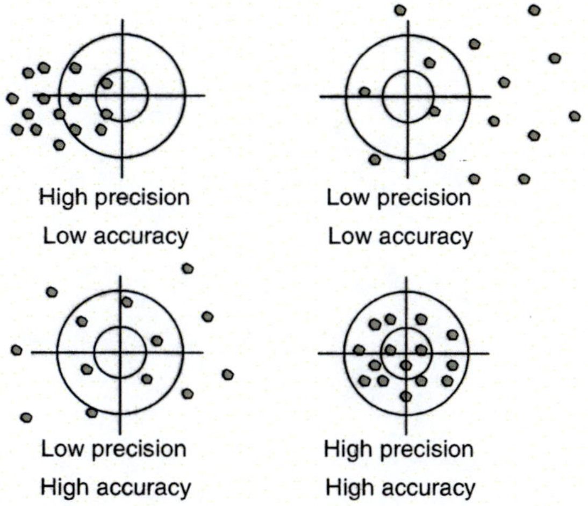

Fig. 3.3 The four extreme combinations of uncertainty that can occur with a sampling plan

of material inspected (laboratory sample[1] size, test portion size, and number of aliquots quantified) usually reduces variability. When describing the uncertainty of a process, one must consider the various combinations of accuracy and precision that may occur. As shown in Fig. 3.3, there are four extreme combinations of accuracy and precision: low precision and low accuracy, low precision and high accuracy, high precision and low accuracy, and high precision and high accuracy.

The worst possible situation is to have a mycotoxin test procedure with low precision and low accuracy. The best possible situation is to have a process that has both high precision and high accuracy. The goal associated with detecting a mycotoxin in a bulk shipment is to design a mycotoxin test procedure or sampling plan that has both high precision and high accuracy.

[1]Laboratory sample: smallest quantity of material comminuted in a mill. The laboratory sample may be a portion of or the entire aggregate sample. The laboratory sample should be finely ground and mixed thoroughly using a process that approaches as complete a homogenisation as possible (Food and Agriculture Organization 2001b).

Chapter 4
Sample Selection

Procedures used to take a sample from a bulk lot are extremely important. Every individual item in the lot should have an equal chance of being chosen (called random sampling). Biases are introduced by sample selection methods if equipment and procedures used to select the sample prohibit or reduce the chances of any item in the lot from being chosen. Examples of bias in the sample selection process, shown in Fig. 4.1, are illustrated with the use of a sampling probe that doesn't allow larger particles into the probe, a probe that doesn't reach every location in the shipment, and use of a single probing point in a poorly mixed lot.

If the lot has been blended thoroughly from the various material handling operations, then the contaminated particles are probably distributed uniformly throughout the lot (William 1991). In this situation, the location within the lot from which the sample is drawn is probably not too important. However, if the lot is contaminated because of moisture leaks that cause high moisture clumps or for other localized reasons, then the mycotoxin-contaminated particles may be located in isolated pockets in the lot (Shotwell et al. 1975). If the sample is drawn from a single location, the contaminated particles may be missed or too many contaminated particles may be collected (Fig. 4.2).

Because contaminated particles may not be distributed uniformly throughout the lot, the sample should be an accumulation of many small portions (called incremental samples[1]) taken from many different locations throughout the lot (Bauwin and Ryan 1982; Hurburgh and Bern 1983). For aflatoxins, FAO/WHO recommends that each incremental sample be about 200 g and one incremental portion be taken for every 200 kg of product (Food and Agriculture Organization/World Health Organization 2001). The accumulation of many small incremental portions is called an aggregate sample. If the aggregate sample is larger than desired, the aggregate sample should be blended and subdivided until the desired laboratory sample size is achieved (Fig. 4.3).

[1]Incremental samples: quantity of material taken from a single random place in the lot (Food and Agriculture Organization 2001b).

T.B. Whitaker et al., *Sampling Procedures to Detect Mycotoxins in Agricultural Commodities*,
DOI 10.1007/978-90-481-9634-0_4,

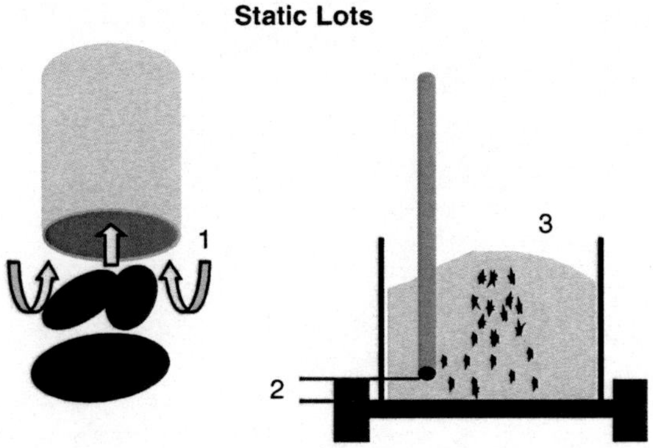

Fig. 4.1 Different types of biases associated with selecting samples from bulk lots. (1) Particles larger than probe opening; (2) some particles in the lot can't be reached; (3) using a single probing point with an unmixed lot

Fig. 4.2 Multiple probing points should be used with a non-homogeneous lot

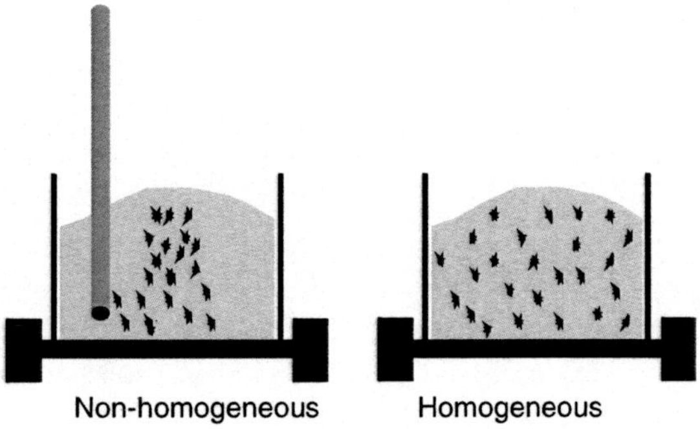

Fig. 4.3 A laboratory sample is removed from an aggregate sample. An aggregate sample is the accumulation of many small incremental samples taken from many different locations in the lot

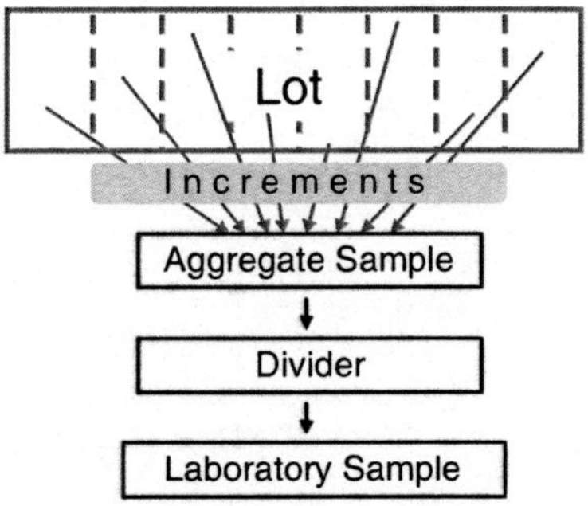

The smallest sample size that is subdivided from the aggregate sample and comminuted in a grinder in the sample preparation step is called the laboratory sample. It is generally more difficult to obtain a representative (lack of bias) laboratory sample from a lot at rest (static lot) than from a moving stream of the product (dynamic lot) as the lot is moved from one location to another. Sample selection methods differ depending on whether the lot is static or dynamic.

4.1 Static Lots

A static lot can be defined as a large mass of materials contained either in a single large container such as a wagon, truck, or railcar or in many small containers such as sacks or boxes, and the materials are stationary at the time a sample is selected (Food and Agriculture Organization 2001b). When drawing an aggregate sample from a bulk container, a probing pattern should be developed so that product can be collected from different locations in the lot. An example of several probing patterns used by the USDA to collect aggregate samples from peanut lots is shown in Fig. 4.4 (United States Department of Agriculture 1975; Parker et al. 1982; Whitaker and Dowell 1995), where the insertion positions for the probe are marked by an 'x' for a five-point sampling pattern, with additional positions indicated by an '0' to give an eight-point sampling pattern. The sampling probe should be long enough to reach the bottom of the container when possible, should not restrict any item in the lot from being selected, and should not alter the items in the lot.

Attempts should be made using a sampling rate similar to the 200 g of incremental sample per 200 kg product as mentioned above. However, it may not be possible to

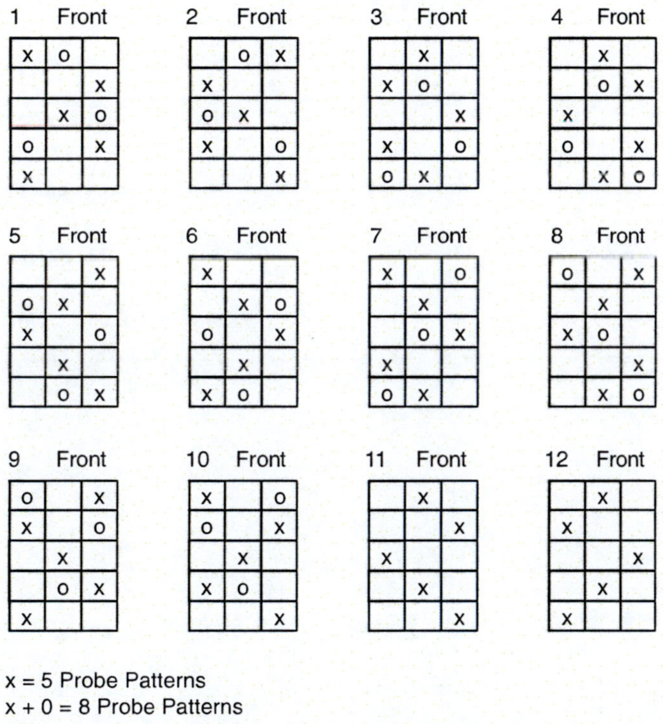

Fig. 4.4 Example of several five- and eight-probe patterns used by the U.S. Department of Agriculture to sample peanuts for grade

achieve the suggested sampling rate because of the design of the sampling equipment, size of the individual containers, and the size of the lot. As an example, assume that a laboratory sample (TSS) of 5 kg is to be taken from a lot (LS) of 25,000 kg. The preferred incremental sample size (ISS) is 0.200 kg. The minimum number of increments needed to provide a laboratory sample of 5 kg is equal to TSS/ISS= 25, this means 25 incremental portions of 200 g each.

Now if only 25 incremental portions are to be taken from the lot of 25,000 kg, this corresponds to a sampling rate of 200 g incremental sample every 1,000 kg (= 25,000 kg/25) of product. This value is much larger than the recommended sampling rate of 1 incremental portion every 200 kg of product and therefore "not acceptable". A sampling rate of 200 g of incremental sample per 200 kg product corresponds to the withdrawal of 125 incremental samples (= 25,000 kg/200 kg) of 200 g each, amounting to an aggregate sample of 25 kg (five times bigger than the needed laboratory sample size of 5 kg). To comply with the recommendations, therefore, a 25 kg aggregate sample must be taken, then representatively subdivided to obtain the desired amount of laboratory sample (5 kg).

In general if the accumulated aggregate sample is larger than required, the aggregate sample should be thoroughly blended and reduced to the required laboratory sample size using a suitable divider that randomly removes a laboratory sample from the aggregate sample. A flow diagram showing the interactions between all the variables is shown in Fig. 4.5.

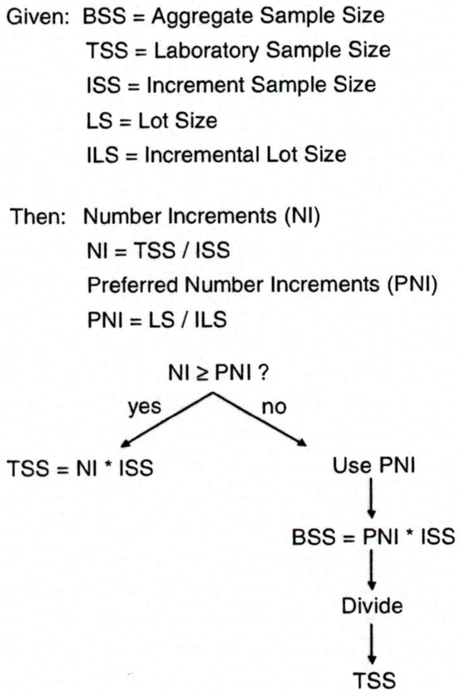

Fig. 4.5 Interaction between lot size, increment size, and laboratory sample size

When sampling a static lot in separate containers such as sacks or retail containers, the sample should be taken from many containers dispersed throughout the lot. When storing sacks in a storage facility, access lanes should be constructed in order to allow access to sacks at interior locations. The recommended number of containers sampled can vary from one in four in small lots (less than 20 metric tonnes) to the square root of the total number of containers for large (greater than 20 metric tonnes) lots (Food and Agriculture Organization/World Health Organization 2001). If the lot is in a container where access is limited, the incremental sample should be drawn when the product is either being removed from or being placed into the container.

4.2 Dynamic Lots

True random sampling can be more nearly achieved when selecting an aggregate sample from a moving stream as the product is transferred (i.e. conveyor belt) from one location to another.

When sampling from a moving stream, small incremental samples of product should be taken along the entire length of, and across the entire cross section of, the moving stream (Fig. 4.6). All the incremental samples should be combined to obtain the aggregate sample; and if the aggregate sample is larger than required, it should be blended and subdivided to obtain the desired size of laboratory sample.

Automatic sampling equipment such as cross-cut samplers (Fig. 4.7) are commercially available with timers that automatically pass a diverter cup through the moving stream at predetermined and uniform intervals. When automatic equipment is not available, a person can be assigned to manually pass a cup though the stream at periodic intervals to collect incremental samples. Whether using automatic or manual methods, small increments of product should be collected and composited at frequent and uniform intervals throughout the entire time product flows past the sampling point.

Cross-cut samplers should be installed in the following manner: (a) the plane of the opening of the sampling cup should be perpendicular to the direction of flow; (b) the sampling cup should pass through the entire cross sectional area of the stream; and (c) the opening of the sampling cup should be wide enough to accept all items of interest in the lot. As a general rule, the width of the sampling cup opening should be two to three times the largest dimensions of the items in the lot.

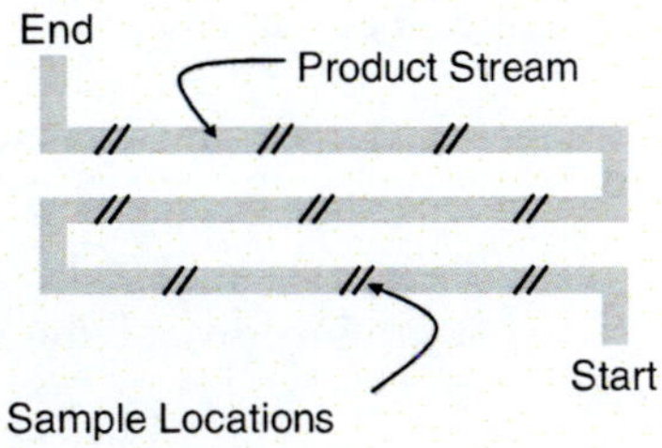

Fig. 4.6 Sample selection from a moving stream of product should be the accumulation of many small incremental samples taken from the beginning to the end of the product stream

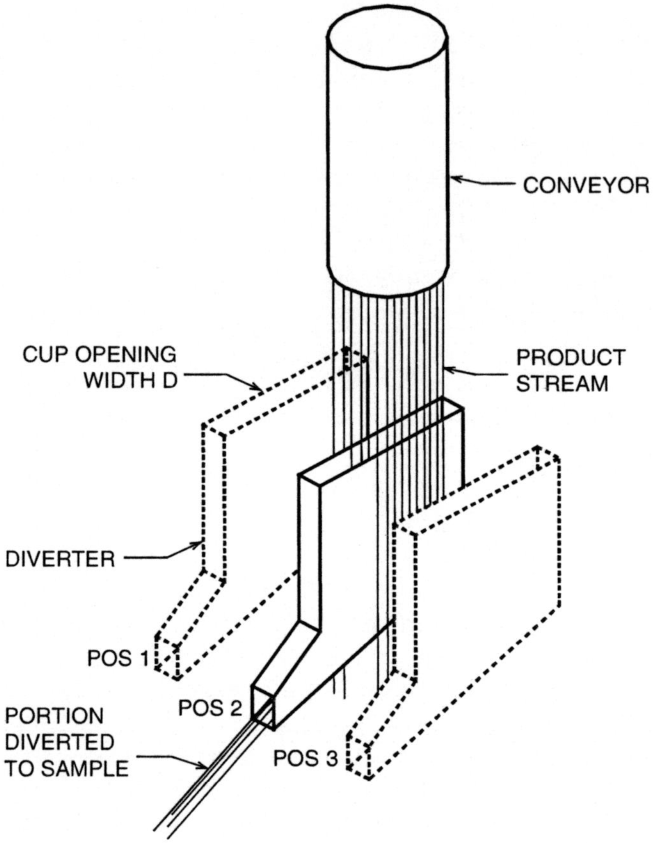

Fig. 4.7 The automatic sampler cup should move at a constant velocity and cut through the entire stream of product

The size of the aggregate sample, S in kg, taken from a lot by a cross cut sampler is

$$S = (D)(L)/(T)(V), \tag{4.1}$$

where D is the width of the sampling cup opening in cm, L is the lot size in kg, T is interval or time between cup movement through the stream in seconds, and V is cup velocity in cm/s.

Equation 4.1 can also be used to compute other terms of interest such as the time between cuts, T. For example, the required time, T, between cuts of the sampling cup to obtain a 10 kg aggregate sample from a 30,000 kg lot where the sampling cup width is 5.08 cm (2 inches), and the cup velocity through the stream is 30 cm/s, can be determined by solving for T in Eq. 4.1.

$$\mathbf{T = (5.08\,cm \times 30{,}000\,kg)/(10\,kg \times 30\,cm\,/\,s) = 508\,s}$$

Table 4.1 Product sample sizes used by the United States Food and Drug Administration (FDA)

Product	Description	Package Type	Lot Size	Number of Incremental Samples	Incremental Sample Size (lbs)	Laboratory Sample (lbs)[a]
Peanut butter	Smooth	Consumer bulk		24	0.5	12
				12	1	12
Peanuts	Crunchy butter, raw, roasted, ground topping	Consumer & bulk		48	1	48
Tree nuts	In shell, shelled, slices or flour paste	Consumer & bulk		10	1	10
				50	1	50
				12	1	12
Brazil nuts	Inshell in import status	Bulk	<200 bags	20	1	
			201–800 bags	40	1	
			801–2,000 bags	60	1	
Pistachio nuts	inshell in import status	Bulk	75,000 lbs	20% of units		50 lbs
						25 lbs
			<75,000 lbs	20% of units		
Corn	Shelled, meal, flour, grits	Consumer & bulk		10	1	10
Cottonseed		Bulk		15	4	60
Oilseed meals	Peanut, cottonseed	Bulk		20	1	20
Edible seeds	Pumpkin, melon, sesame, etc.	Bulk		50	1	50
Ginger root	Dried, whole ground	Consumer & bulk		??		15
			"n" units	10	10 × 0.06	10
Milk	Whole, low fat, skim	Bulk consumer				10
				10	1	10
Small grains	Sorghum, wheat, barley, etc.	Bulk		10	1	10
Dried fruit	i.e. figs	Consumer & bulk		50		50
Mixtures	– Commodity particles large	Consumer & bulk		50		50
				10		10
	– Commodity particle finely ground					

[a] 1 lb corresponds to 0.454 kg

If the lot is moving at 1,000 kg/min, the entire lot will pass through the sampler in 30 min and only three or four cuts will be made by the cup through the lot. This may be considered too infrequent, because too much product passes the sampling point between the times the cup cuts through the stream. The interaction among the variables in Eq. 4.1 needs to be fully understood in terms of the amount of aggregate sample accumulated and the frequency of taking product.

4.3 Aggregate Versus Laboratory Sample

Because contaminated particles may not be uniformly dispersed throughout the lot, many incremental samples are taken from many different locations throughout the lot and combined to form an aggregate sample. As a result, the aggregate sample is usually larger than the desired laboratory sample size used to estimate the lot mycotoxin concentration.

- For granular materials, the laboratory sample is the amount of granular product ground in a mill in the sample preparation step.
- For finely ground materials (corn flour) or liquids (milk), the laboratory sample is easily subdivided to obtain the test portion, i.e. the amount of sample used in the analytical step to quantify the mycotoxin.

When the aggregate sample is larger than the laboratory sample, dividers should be used to withdraw the desired laboratory sample amount. Various types of sample divider may be used, including mechanical devices such as a Boerner or riffle divider (Parker et al. 1982). These devices are considered to give random divisions, and in this case the aggregate sample does not have to be blended before the laboratory sample is removed. However, if the laboratory sample is to be removed from the aggregate sample using quartering or a manual device such as a cup or scoop, then the aggregate sample should be blended before the laboratory sample is removed.

As will be shown later, the size of the laboratory sample put through the grinder should be as large as possible. As the laboratory sample becomes smaller, the uncertainty (precision) associated with estimating the true lot mycotoxin concentration becomes greater. Recommended laboratory sample sizes for various commodities are shown in Table 4.1.

Chapter 5
Sample Preparation

Once an aggregate sample has been collected and the laboratory sample withdrawn (if the aggregate sample is larger than the required laboratory sample), the laboratory sample must be prepared for mycotoxin quantification. Since it is not practical to extract the mycotoxin from a large laboratory sample, the mycotoxin is usually extracted from a much smaller portion of product (test portion) taken from the comminuted laboratory sample. If the commodity is a granular product such as shelled corn, it is essential that the entire laboratory sample be comminuted in a suitable mill and thoroughly homogenized before a test portion is removed from it (Dickens and Whitaker 1982; Campbell et al. 1986). As stated above, it is necessary to grind/comminute the entire laboratory sample before a test portion can be taken for analysis. Figure 5.1 emphasizes the need to comminute a laboratory sample to be able to obtain a smaller particle size and therefore a more homogeneous distribution of the mycotoxins.

Grinders should be used that reduce the particle size of the laboratory sample to the smallest size possible. Grinders that produce small particles produce a more homogeneous laboratory sample with respect to the mycotoxin distribution (Fig. 5.1). As a result the mycotoxin concentration in the test portion will more nearly reflect the true mycotoxin concentration of the laboratory sample.

Some grinders, such as the Romer mill (Malone 2000) and the USDA peanut mill (Dickens and Satterwhite 1969), are designed to automatically subsample the laboratory sample during the grinding process and provide a comminuted test portion after the grinding process. If the mill does not provide such a feature, the test portion can be obtained using a riffle divider. If the test portion is obtained using a manual device such as a scoop, the comminuted laboratory sample must be blended before scooping out a test portion.

Usually there is no sample preparation step associated with laboratory samples of non-granular products such as liquids (milk) or paste (peanut butter). A small portion of the aggregate sample may have to be removed for mycotoxin analysis because the entire aggregate sample cannot be analyzed. However, it is important to blend or mix liquid samples and paste samples before removing a small portion for mycotoxin analysis.

T.B. Whitaker et al., *Sampling Procedures to Detect Mycotoxins in Agricultural Commodities*,
DOI 10.1007/978-90-481-9634-0_5,

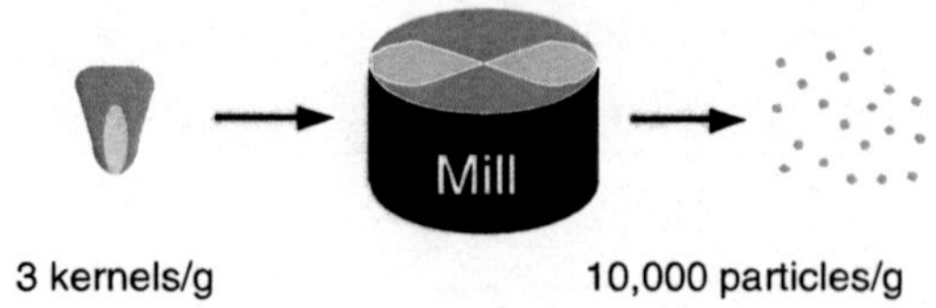

Fig. 5.1 A laboratory sample of granular product (3 kernels/g) should be ground in a mill to reduce particle size (10,000 particles/g)

Test portion sizes vary, but usually are of the order of 20–1,000 g, depending on the particle size. The smaller the particle size, the smaller the test portion can be without increasing error or uncertainty. For most analytical methods, a test portion of 20–50 g is recommended. Analysis of test portions less than 20 g is likely to increase the uncertainty of the result.

Chapter 6
Analytical Quantification

Once the test portion is removed from the ground laboratory sample, the mycotoxin is extracted by blending a solvent with the comminuted test portion. Before the mycotoxin can be quantified in the solvent extract, analytical methods usually consist of several steps related to removing interfering compounds (i.e. oils) and concentrating the mycotoxins for quantification. These steps include centrifugation, filtration, drying, and dilution (Dickens and Satterwhite 1969; Association of Official Analytical Chemists 1990; Nesheim 1979; Steyn et al. 1991).

There are several different analytical methods that can be used to quantify the mycotoxin extracted from the test portion. Three examples are: thin layer chromatography, immunoassay methods such as enzyme-linked immunosorbent assay (ELISA) that use antibody technology, and high performance liquid chromatography (HPLC) (Steyn et al. 1991). Organizations such as the Association of Official Analytical Chemists (AOAC) evaluate the performance of analytical methods using collaborative studies.

There are several sources of biases associated with analytical methods. For example, less than 100% of the mycotoxin may be extracted from the test portion by the solvents; compounds other than mycotoxins may be extracted into the solvent and mistakenly quantified as a mycotoxin; mycotoxin standards used in quantification may not be exact; and instruments to measure the mycotoxin may not be correctly calibrated.

T.B. Whitaker et al., *Sampling Procedures to Detect Mycotoxins in Agricultural Commodities*, DOI 10.1007/978-90-481-9634-0_6, © Springer Science+Business Media B.V. 2010

Chapter 7
Accept/Reject Limit

Once the mycotoxin concentration is quantified, the concentration value is used to estimate the true lot concentration or is compared to an accept/reject limit (ARL). The ARL is a predefined threshold concentration, usually equal to a legal limit used in regulatory applications. If the mycotoxin concentration in a test portion taken from a laboratory sample is less than or equal to the ARL, the lot is accepted, otherwise the lot is rejected.

When lots are inspected by regulatory agencies, the ARL is usually set equal to the legal limit. However, manufacturers of consumer-ready products will often use an ARL less than the legal limit to reduce the chances that consumer-ready products will be found by regulatory agencies with mycotoxin concentrations above the legal limit. Often private industry will use an ARL that is about half the legal limit (personal communications).

Many countries agree on the need to establish legal limits, but often disagree on the value of the limit. A survey by FAO in 2003 (Food and Agriculture Organization 2003) showed that some countries have aflatoxin legal limits based upon B_1 only and some countries use total ($B_1 + B_2 + G_1 + G_2$) aflatoxin and these legal limits vary widely. The CODEX Committee on Food Additives and Contaminants has established a standard aflatoxin limit for peanuts at 15 ng/g total aflatoxin for raw peanuts destined for further processing traded on the international market (Food and Agriculture Organization 2001a). This limit does not infringe on any nations' internal limits.

T.B. Whitaker et al., *Sampling Procedures to Detect Mycotoxins in Agricultural Commodities*, DOI 10.1007/978-90-481-9634-0_7, © Springer Science+Business Media B.V. 2010

Chapter 8
Random Variation

Even when using accepted sampling, sample preparation, and analytical procedures (Campbell et al. 1986; Malone 2000; Dickens and Satterwhite 1969; Association of Official Analytical Chemists 1990; Nesheim 1979; Steyn et al. 1991), there are errors (the term error will be used to denote variability) associated with each of the steps of the mycotoxin test procedure (Whitaker et al. 1972, 1974, 1976, 1979, 1993, 1998; Dickens et al. 1979; Remington and Schrok 1970; Schatzki 1995a, b). Because of these errors, the true mycotoxin concentration in the lot cannot be determined with 100% certainty by measuring the mycotoxin concentration in a test portion taken from a laboratory sample taken from the lot. For example, 10 replicated aflatoxin test results from each of six contaminated shelled peanut lots are shown in Table 8.1 (Whitaker et al. 1972). For each test result in the table, the mycotoxin test procedure consisted of (a) comminuting a 5.45-kg laboratory sample of peanut kernels in a USDA subsampling mill developed by Dickens and Satterwhite (1969), (b) removing a 280-g test portion from the comminuted laboratory sample, (c) solvent extracting aflatoxins from a 280-g test portion as described by AOAC Method II (Association of Official Analytical Chemists 1990), and (d) quantifying the aflatoxins densitometrically using thin layer chromatography (TLC). The 10 aflatoxin test results from each lot are ranked from low to high to demonstrate several important characteristics about replicated aflatoxin test results taken from the same contaminated lot.

First, the wide range among the 10 laboratory sample test results from the same lot reflects the large variability associated with estimating the true mycotoxin content of a bulk lot. In Table 8.1, the variability is described by the standard deviations (SD) and the coefficients of variation (CV). The maximum laboratory sample test result can be four to five times the lot concentration (the average of the 10 laboratory sample test results is the best estimate of the lot concentration). Secondly, the amount of variation among the 10 laboratory sample test results appears to be a function of the lot concentration. In Table 8.1 the best estimate of the lot concentration is the mean or the average of the 10 laboratory sample test results. As the lot concentration (mean) increases, the standard deviation among laboratory sample test results increases, but the standard deviation relative to the lot mean, as measured by the CV, decreases. Thirdly, the distribution of the 10 laboratory

T.B. Whitaker et al., *Sampling Procedures to Detect Mycotoxins in Agricultural Commodities*,
DOI 10.1007/978-90-481-9634-0_8,

Table 8.1 Distribution of aflatoxin test results for ten 5.4 kg laboratory samples from each of six lots of shelled peanuts[a, b]

Lot number	10 Laboratory sample test results (ng/g)										Mean (ng/g)	SD[c] (ng/g)	CV[d] (%)
1	0	0	0	0	2	4	8	14	28	43	10	15	150
2	0	0	0	0	3	13	19	41	43	69	19	24	126
3	0	6	6	8	10	50	60	62	66	130	40	42	105
4	5	12	56	66	70	92	98	132	141	164	84	53	63
5	18	50	53	72	82	108	112	127	182	191	100	56	56
6	29	37	41	71	95	117	168	174	183	197	111	66	59

[a]From Whitaker et al. (1972)
[b]Aflatoxin test results (ng/g)
[c]SD = Standard deviation
[d]CV = Coefficient of variation = SD*100/mean

sample test results for each lot in Table 8.1 is not always symmetrical about the lot concentration.

The distributions among laboratory sample test results are positively skewed, meaning that more than half of the laboratory sample test results are below the true lot concentration. However, the distribution of laboratory sample test results becomes more symmetrical as the lot concentration increases. This skewness can be observed by counting the number of laboratory sample test results above and below the lot concentration in Table 8.1 (average of the 10 laboratory sample test results).

If a single laboratory sample from a contaminated lot is tested for aflatoxin, there is more than a 50% chance that the laboratory sample test result will be lower than the true lot concentration. While it cannot be shown in Table 8.1, the skewness is greater for small sample sizes and the distribution becomes more symmetrical as laboratory sample size increases (Whitaker et al. 1979). The above characteristics described by Table 8.1 for aflatoxin in shelled peanuts are also generally found for other mycotoxins and other commodities (Whitaker et al. 1993, 1998, 2000; Hart and Schabenberger 1998; Johansson et al. 2000a; Cucullu et al. 1986).

The sources of the variability among mycotoxin test results in Table 8.1 are associated with each step of the mycotoxin test procedure (Fig. 2.2). The sampling, sample preparation, and analytical steps of the mycotoxin test procedure each contribute to the total variability observed among mycotoxin test results shown in Table 8.1.

As shown in Fig. 8.1, the total error or variability is the sum of the sampling, sample preparation, and analytical variability.

Among the statistical measures of variability shown in Eq. 3.2 to 3.4, only the variance is additive. Therefore, it is assumed that the total variance (VT) associated with a mycotoxin test procedure is the sum of the sampling variance (VS), sample preparation variance (VSP), and analytical variance (VA).

$$\mathrm{VT} = \mathrm{VS} + \mathrm{VSP} + \mathrm{VA} \tag{8.1}$$

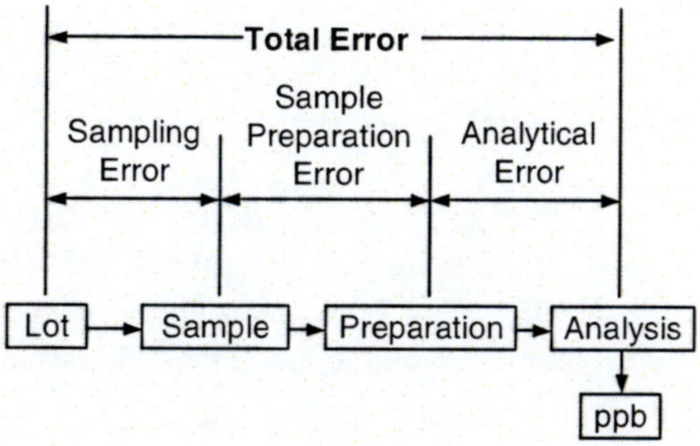

Fig. 8.1 Total error of the mycotoxin-test procedure is the sum of sampling, sample preparation, and analytical errors.

Reasons why each step of the mycotoxin test procedure contributes to the overall variability are discussed below. An example of the magnitude of the contribution of each step of the mycotoxin test procedure to the total variability is shown for measuring aflatoxin in shelled corn (maize). Another example of uncertainty associated with sampling maize for fumonisin is described in Appendix A.

8.1 Sampling Variability

Studies by researchers on a wide variety of agricultural products (peanuts, cottonseed, shelled corn and pistachio nuts) indicate that, especially for small laboratory sample sizes, the sampling step is usually the largest source of variability associated with the mycotoxin test procedure (Whitaker et al. 1972, 1974, 1976, 1979, 1993; Dickens et al. 1979; Remington and Schrok 1970; Schatzki 1995a, b). Even when using accepted sample selection equipment and random sample selection procedures, sampling error is inevitably large because of the distribution among contaminated particles within a lot. Studies by researchers on a wide variety of agricultural products such as peanuts and shelled corn (Cucullu et al. 1977, 1986; Shotwell et al. 1974) indicate that a very small percentage (0.1%) of the kernels in the lot are contaminated and the concentration on a single kernel may be extremely high. Cucullu et al. (1986) reported aflatoxin concentrations in excess of 1,000,000 ng/g (parts per billion, ppb) for individual peanut kernels and 5,000,000 ng/g for cottonseed. Shotwell et al. (1974) reported finding over 400,000 ng/g of aflatoxin in a corn kernel.

Because of this extreme range in aflatoxin concentrations among a few contaminated kernels in a lot, variation among replicated laboratory sample test results tends to be large. As an example, the sampling variance, VS, associated with testing shelled corn was estimated empirically (Johansson et al. 2000a) and is shown in Eq. 8.2 for any sample size, ns.

$$\mathbf{VS} = (\mathbf{12.95} / \mathbf{ns})\mathbf{M}^{0.98} \tag{8.2}$$

where M is the aflatoxin concentration in the lot in nanograms of total aflatoxin per g of corn (ng/g) or parts per billion (ppb), ns is the mass of shelled corn in the laboratory sample in kg (kernel count per gram was 3.0). From Eq. 8.2 one can see that the sampling variance is a function of the aflatoxin concentration, M, and sample size, ns. For example, the sampling variance among replicated 0.91 kg (2 lb) samples taken from a lot of shelled corn at 20 ng/g is 268.1 and the coefficient of variation is 81.8%.

Researchers have developed equations to describe the sampling variance for several commodities and mycotoxins (Whitaker et al. 1972, 1974, 1976, 1979, 1993, 1998, 2000; Dickens et al. 1979; Remington and Schrok 1970; Schatzki 1995a, b; Hart and Schabenberger 1998; Johansson et al. 2000a). The equations are specific for the type of mycotoxin and the type of product studied, but generally show that sampling variance is a function of concentration, increases with an increase in concentration, and decreases with an increase in laboratory sample size.

8.2 Sample Preparation Variability

Once the laboratory sample has been taken from the lot, the sample must be prepared for mycotoxin quantification. Since it is not practical to extract the mycotoxin from a large laboratory sample, this is comminuted in a mill and the mycotoxin is extracted from a small test portion taken from the comminuted laboratory sample. If the commodity is a granular product such as shelled corn, it is essential that the entire laboratory sample be comminuted in a suitable mill before a test portion is removed from the laboratory sample (Campbell et al. 1986). Removing a test portion of whole seed from the laboratory sample before the comminuting process is simply a sample size reduction process and eliminates the benefits associated with the larger size laboratory sample of granular product.

After the laboratory sample has been comminuted in a mill to reduce particle size, a test portion is removed for mycotoxin extraction. It is assumed that the mycotoxin distribution among contaminated particles in the comminuted laboratory sample is similar to the distribution among contaminated kernels found in the laboratory sample before comminution. As a result, there is also variability among replicated test portions taken from the same comminuted laboratory sample. The sample preparation variance is not as large as the sampling variance due to the large number of comminuted particles in the test portion, but may still be significant (Whitaker et al. 1974; Johansson et al. 2000a; Maestroni et al. 2005). An example of sample preparation variance for aflatoxin and shelled corn, VSP, is shown below in Eq. 8.3 for any test portion size nss (Johansson et al. 2000a).

$$\mathbf{VSP = (62.70 / nss)M^{1.27}} \tag{8.3}$$

where M is the aflatoxin concentration in the laboratory sample in ng/g, nss is the mass of shelled corn in the test portion in grams taken from the comminuted laboratory sample. The variance in Eq. 8.3 also reflects the use of a Romer mill that produces a particle size where most of the particles will pass through a number 20 screen. From Eq. 8.3, it can be seen that the sample preparation variance is also a function of the aflatoxin concentration (M) and the size of the test portion (nss). The sample preparation variance associated with a 50 g test portion taken from a laboratory sample at 20 ng/g is 56.3 and the CV is 37.5%.

Researchers have developed equations to describe the sample preparation variance for several commodities, mill types, and mycotoxins (Whitaker et al. 1972, 1974, 1976, 1979, 1993; Dickens et al. 1979; Remington and Schrok 1970;

Schatzki 1995a, b). The equations are specific for the mycotoxin, mill type (particle size), and the type of product used in the study. The type mill type affects the particle size distribution. If the average particle size decreases (number of particles per unit mass increases), then the sample preparation variance for a given size test portion decreases.

8.3 Analytical Variability

Once the test portion is removed from the comminuted laboratory sample, the mycotoxin is solvent extracted. Analytical methods usually involve several steps such as solvent extraction, centrifugation, drying, dilution, and quantification (Nesheim and Trucksess 1986). As a result, there can be considerable variation among replicated analyses on the same test portion extract. The analytical variance (VAh) associated with high performance liquid chromatography (HPLC) techniques used to measure aflatoxin in a test portion taken from a comminuted laboratory sample of shelled corn is given by Eq. 8.4 (Johansson et al. 2000a) for any number of aliquots.

$$\mathbf{VAh} = (\mathbf{0.143} / \mathbf{na})\,\mathbf{M}^{1.16} \tag{8.4}$$

where M is the aflatoxin concentration in the test portion in ng/g, na is the number of aliquots quantified by HPLC methods. Applying this equation, typical values for analytical variance and CV associated with using HPLC to measure aflatoxin in a comminuted subsample of corn at 20 ng/g, are 4.6 and 10.7%, respectively.

High performance liquid chromatography tends to have less variability than other analytical technologies such as thin layer chromatography (TLC) and immunoassay (ELISA) methods (Whitaker et al. 1996). Using precision estimates from collaborative studies, the analytical variances associated with TLC (VAt) and ELISA (VAe) methods to measure aflatoxin in shelled corn are shown in Eqs. 8.5 and 8.6, respectively.

$$\mathbf{VAt} = (\mathbf{0.316} / \mathbf{na})\,\mathbf{M}^{1.744} \tag{8.5}$$

$$\mathbf{VAe} = (\mathbf{0.631} / \mathbf{na})\,\mathbf{M}^{1.293} \tag{8.6}$$

Typical coefficients of variation associated with measuring aflatoxin in a test portion at 20 ng/g with the TLC and ELISA methods are 38.3 and 27.5%, respectively. The variability or CV associated with HPLC (10.7%, Eq. 8.4) is lower than either TLC or ELISA.

All of the analytical variance information described above reflects results from single laboratories and does not reflect among-laboratory variances. As a result, some laboratories may have higher or lower variances than those reported in Eqs. 8.4–8.6. Among-laboratory variance is about double the within-laboratory variance (Whitaker et al. 1996).

8.4 Total Variability

As shown in Fig. 8.3 and Eq. 8.1, the total variability, VT (using variance as the statistical measure of variability) associated with a mycotoxin test procedure is equal to the sum of the sampling (VS), sample preparation (VSP), and analytical (VA) variances associated with each step of the mycotoxin test procedure. The total variability associated with testing shelled corn for aflatoxin, grinding the laboratory sample in a Romer mill, and quantifying aflatoxin in a test portion by immunoassay is the sum of Eqs. 8.2, 8.3, and 8.6.

$$\mathbf{VT} = (12.95/\mathbf{ns})\,\mathbf{M}^{0.98} + (62.70/\mathbf{nss})\,\mathbf{M}^{1.27} + (0.631/\mathbf{na})\,\mathbf{M}^{1.293} \tag{8.7}$$

Using Eq. 8.7, the total, sampling, sample preparation, and analytical variances associated with testing shelled corn lots over a range of aflatoxin concentrations (M) when using a 0.91-kg (1 lb) laboratory sample (ns), grinding the laboratory sample in a Romer mill, taking a 50-g test portion (nss) from the comminuted laboratory sample, and quantifying aflatoxin in one aliquot (na) by immunoassay methods are shown in Fig. 8.2. The CV associated with each step is shown in Fig. 8.3.

When sampling a bulk shipment of shelled corn for aflatoxin at 20 ng/g, the magnitude of the variance associated with each step (defined in Table 8.2) of the above aflatoxin test procedure (Eq. 8.7) are shown below in Eq. 8.8.

$$\mathbf{VT} = 268.1 + 56.3 + 30.4 = 354.8 \tag{8.8}$$

As shown in Table 8.2, the sampling, sample preparation, and analytical variances account for 75.5%, 15.9%, and 8.6% of the total mycotoxin testing variance, respectively.

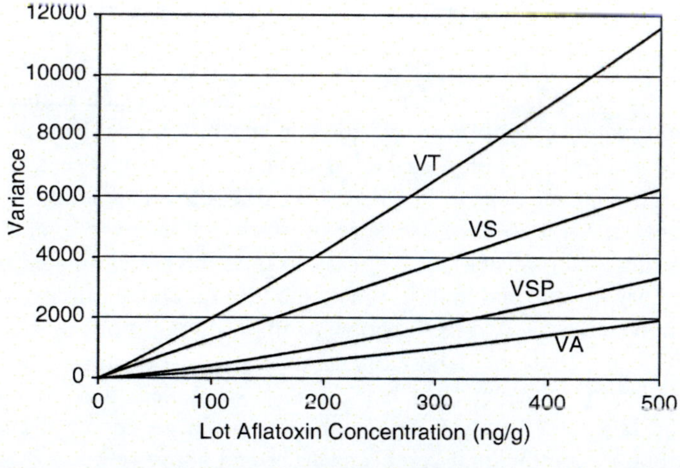

Fig. 8.2 Variability of each step of the aflatoxin-test procedure, as measured by the variance (V), increases with aflatoxin concentration. The total variance, VT, is the sum of sampling variance, VS, sample preparation variance, VSP, and analytical variance, VA

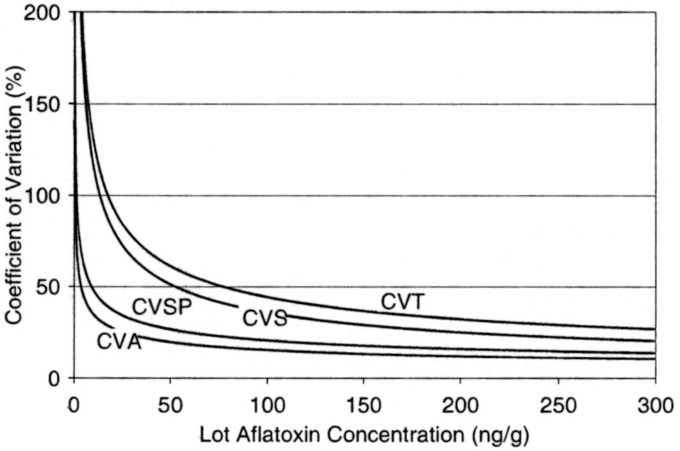

Fig. 8.3 Variability of each step of the aflatoxin-test procedure, as measured by the coefficient of variation (CV), decreases with aflatoxin concentration. The total CV (CVT) is not the sum of the sampling CV (CVS), sample preparation CV (CVSP), and analytical CV (CVA)

Table 8.2 The variability, as measured by the variance, associated with 0.91 kg laboratory sample, Romer mill to grind the laboratory sample, a 50 g test portion, measuring aflatoxin in 1 aliquot by immunoassay analytical methods to measure aflatoxin in shelled corn at 20 ng/g. Sampling, sample preparation, and analysis errors accounts for about 75.5%, 15.9%, and 8.6% of the total error, respectively

Lot shelled corn at 20 ng/g aflatoxin		
Test procedure	Variance	Variance ratio[a] (%)
Laboratory sample – 0.91 kg	268.1	75.5
Sample preparation – Romer mill, 50 g test portion	56.3	15.9
Immunoassay analytical method, analysis of 1 aliquot	30.4	8.6
Total	354.8	100.0

[a]Variance ratio = Variance associated with each step of test procedure divided by the total variance

As the above example demonstrates, the sampling step accounts for most of the variability (uncertainty) associated with the total variability of a mycotoxin test procedure because of the distribution among contaminated seeds in a lot. For shelled corn, it is estimated that only 6 kernels in 10,000 are contaminated in a lot at 20 ng/g (Johansson et al. 2000b). Because of this extreme mycotoxin distribution among seeds in a contaminated lot, it is easy to miss the contaminated seeds with a small laboratory sample and underestimate the true lot concentration.

On the other hand, if the laboratory sample contains one or more highly contaminated seeds, then the laboratory sample can over-estimate the true mycotoxin contamination in the lot. Even using proper sample selection techniques, the variation among laboratory sample concentrations is large due to the mycotoxin distribution among individual kernels as described above.

Chapter 9
Reducing Variability of a Mycotoxin Test Procedure

The only way to achieve a more precise estimate of the true lot concentration is to reduce the total variability of the mycotoxin test procedure. The total variability of the test procedure can be reduced by reducing the variability associated with each step of the mycotoxin test procedure. Increasing the size of the laboratory sample can reduce the sampling variability. The sample preparation variability can be reduced either by increasing the size of the test portion and/or by increasing the degree of comminuting (increasing the number of particles per unit mass in the test portion). The analytical variance can be reduced by either increasing the number of aliquots quantified by the analytical method and/or using a more precise quantification method (e.g. using HPLC instead of TLC). If the variability associated with one or more of these steps can be reduced, then the total variability associated with a mycotoxin test result can be reduced (Eq. 8.1).

Decreasing the total variability (improving precision) associated with a mycotoxin test procedure will decrease the range of possible aflatoxin test results when replicated tests are made on the same lot. The range of mycotoxin test results associated with any size laboratory sample, any size test portion, and number of analyses about the lot concentration M can be estimated from the total variance, VT, or standard deviation, SD, (square root of the total variance) associated with the mycotoxin test procedure. For a large number (>20) of laboratory samples taken from a lot, approximately 95% of all laboratory sample test results will fall in a range between a low of (M – 1.96*SD) and a high of (M + 1.96*SD).

As an example, when sampling a lot of shelled corn at 20 ng/g using a 0.91-kg (2 lb) laboratory sample (ns), grinding the laboratory sample in a Romer mill, taking a 50-g test portion (nss) from the comminuted laboratory sample, and quantifying aflatoxin in one aliquot (na) by immunoassay method, Eq. 8.8 shows that the total variance and standard deviation are 354.8 and 18.8, respectively. The range of aflatoxin test results should fall between 20+/– (1.96*18.8) or 20 +/– 36.9 or 0 and 56.9 ng/g (Table 9.1).

T.B. Whitaker et al., *Sampling Procedures to Detect Mycotoxins in Agricultural Commodities*, DOI 10.1007/978-90-481-9634-0_9, © Springer Science+Business Media B.V. 2010

Table 9.1 A variance of 354.8 associated with the aflatoxin-test procedure described in the table indicates that aflatoxin test results will vary from a low of 0 ng/g to a high of 57 ng/g when sampling a lot of shelled corn contaminated at 20 ng/g

Test procedure	Variance
Sampling – 0.91 kg laboratory sample	268.1
Sample prep – Romer, 50 g test portion	56.3
Analytical method – TLC, 1 aliquot for analysis	30.4
Total	354.8

$$\begin{aligned}\text{Range} &= 20 \pm 1.96\sqrt{354.8}\\ &= 20 \pm 36.9\ \text{ng/g}\\ &= 0\ \text{to}\ 56.9\ \text{ng/g}\end{aligned}$$

The calculated range of aflatoxin test results is only valid for a normal distribution where test results are symmetrical about the true lot concentration. As will be shown later, the distribution among aflatoxin test results is usually skewed, but will approach a symmetrical distribution as laboratory sample size becomes large.

9.1 Laboratory Sample Size

The effect of increasing the laboratory sample size on reducing the total variability and the range of mycotoxin test results when testing a contaminated lot of shelled corn at 20 ng/g aflatoxin is shown in Table 9.2 when increasing sample size from 0.91 (2 lbs) to 4.54 kg (10 lbs).

Increasing laboratory sample size by a factor of 5 from 0.91 to 4.54 kg reduces the sampling variance described in Eq. 8.8 by a factor of 5 from 268.1 to 53.6. The total variance is reduced from 354.8 to 137.8.

$$\mathbf{VT = 53.6 + 56.3 + 27.9 = 137.8} \tag{9.1}$$

The range of aflatoxin test results is reduced from 20 ± 36.9 to 20 ± 23.0 ng/g as the laboratory sample size is increased from 0.91 (2 lbs) to 4.54 kg (10 lbs), respectively.

9.2 Test Portion Size

The effect of increasing the size of the test portion from 50 to 100 g on reducing the sample preparation variance is shown in Table 9.3.

Table 9.2 Effect of increasing laboratory sample size on reducing the sampling variability

Laboratory sample size effect – shelled corn at 20 ng/g			
Test Procedure	Variance	Test Procedure	Variance
Sampling – 0.91 kg laboratory sample	268.1	**4.54 kg**	53.6
Sample Preparation – Romer, 50 g test portion	56.3	**Romer, 50 g**	56.3
Analytical method – TLC, 1 aliquot analysed	27.9	**TLC, 1 aliquot**	27.9
Total	354.8	**Total**	137.8
Range	20 ± 36.9	**Range**	20 ± 23.0

Table 9.3 Effect of increasing test portion size on reducing sample preparation variability

Test portion size effect – shelled corn at 20 ng/g			
Test procedure	Variance	Test procedure	Variance
Sampling – 0.91 kg laboratory sample	268.1	**0.91 kg**	268.1
Sample Prep – Romer, 50 g test portion	56.3	**Romer, 100 g**	28.2
Analytical method:TLC, 1 aliquot analysed	27.9	**TLC, 1 aliquot**	27.9
Total	354.8	**Total**	324.2
Range	20 ± 36.9	**Range**	20 ± 35.3

As the test portion size increases from 50 to 100 g, the sample preparation variance is reduced from 56.3 to 28.2. The total variance is reduced from 354.8 to 324.2.

$$\mathbf{VT = 268.1 + 28.2 + 27.9 = 324.2} \tag{9.2}$$

The range of aflatoxin test results is reduced from 20 ± 36.9 to 20 ± 35.3.

9.3 Number of Aliquots Quantified

The effect of increasing number of aliquots quantified in the analytical step from 1 to 2 on reducing the analytical variance for immunoassay type method is shown in Table 9.4.

As the number of aliquots in increased from 1 to 2, the analytical variance is reduced from 27.9 to 14.0. The total variance is reduced from 354.8 to 338.4.

$$\mathbf{VT = 268.1 + 56.3 + 14.0 = 338.4} \tag{9.3}$$

The range of aflatoxin test results is reduced from 20 ± 36.9 to 20 ± 36.1 ng/g.

There is a different cost associated with reducing the variability of each step of a mycotoxin test procedure. One needs to try and maximize the variance reduction

Table 9.4 Effect of increasing number of aliquots quantified for aflatoxin on reducing analytical variability

Number of aliquots effect – shelled corn at 20 ng/g			
Test procedure	Variance	Test procedure	Variance
Sampling - 0.91 kg laboratory sample	268.1	**0.91 kg**	268.1
Sample prep - Romer, 50 g test portion	56.3	**Romer, 50 g**	56.3
Analytical method: TLC, 1 aliquot analysed	27.9	**TLC, 2 aliquots**	14.0
Total	354.8	**Total**	338.4
Range	20 ± 36.9	**Range**	20 ± 36.1

for a given cost. Increasing laboratory sample size is usually the best use of resources when reducing the total variability of mycotoxin test results.

Chapter 10
Designing Mycotoxin Sampling Plans

Because of the variability among laboratory sample test results, two types of mistakes are associated with any mycotoxin-sampling plan. First, good lots (lots with a concentration less than or equal to the legal limit) will test bad and be rejected by the sampling plan. The chances of making this type of mistake is often called the seller's risk (false positives or type I, α error) since these lots will be rejected at an unnecessary cost to the seller of the product. Secondly, bad lots (lots with a concentration greater than the legal limit) will test good and be accepted by the sampling program. The chances of making this type of mistake is called the buyer's risk (false negatives, or type II, β error) since contaminated lots will be processed into feed or food causing possible health problems and/or economic loss to the buyer of the product.

In order to maintain an effective regulatory and/or quality control program, the above two risks associated with a sample plan design must be evaluated (Fig. 10.1). Based upon these evaluations, the costs and benefits (benefits refers to removal of mycotoxin contaminated lots) associated with a mycotoxin sampling plan need to be evaluated.

A lot is termed bad when the laboratory sample test result X is above some predefined accept/reject limit X_c and the lot is termed good when X is less than or equal to X_c. While X_c is usually equal to the legal limit Mc, X_c can be greater than or less than Mc. For a given sampling plan design, lots with a mycotoxin concentration M will be accepted with a certain probability P(M) (called acceptance probability) = prob($X< X_c$ |M) by the sampling plan. A plot of P(M) versus the lot concentration M is called an operating characteristic (OC) curve. Figure 10.2 depicts the general shape of an OC curve.

As M approaches 0, P(M) approaches 1 or 100%, and as M becomes large, P(M) approaches zero. Lots with little to no contamination (M = 0) are accepted by the sampling plan 100% of the time; lots with very high levels of contamination (M = large) are never accepted (rejected 100% of the time) by the sampling plan; lots with contamination levels near the accept/reject limit are accepted about 50% of the time by the sampling plan. The shape of the OC curve is uniquely defined for a particular sampling plan design with designated values of laboratory sample size, degree of comminution, test portion size, type of analytical method, and number of analyses, and the accept/reject limit X_c.

T.B. Whitaker et al., *Sampling Procedures to Detect Mycotoxins in Agricultural Commodities*,
DOI 10.1007/978-90-481-9634-0_10,

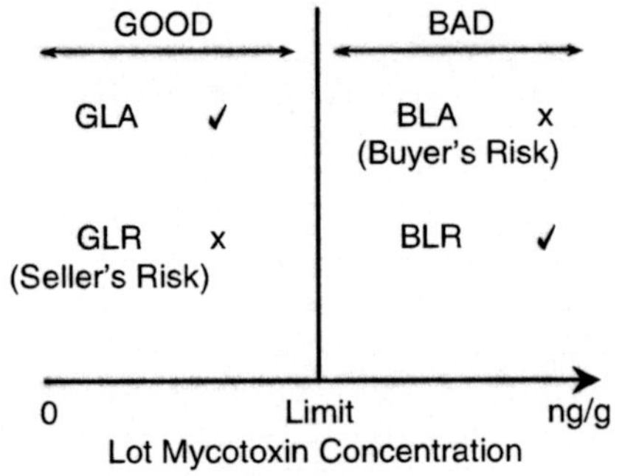

Fig. 10.1 Four possible outcomes when classifying lots as good or bad based upon their mycotoxin concentration. Good lots rejected (GLR) and bad lots accepted (BLA) are incorrect decisions. Good lots accepted (GLA) and bad lots rejected (BLR) are correct decisions

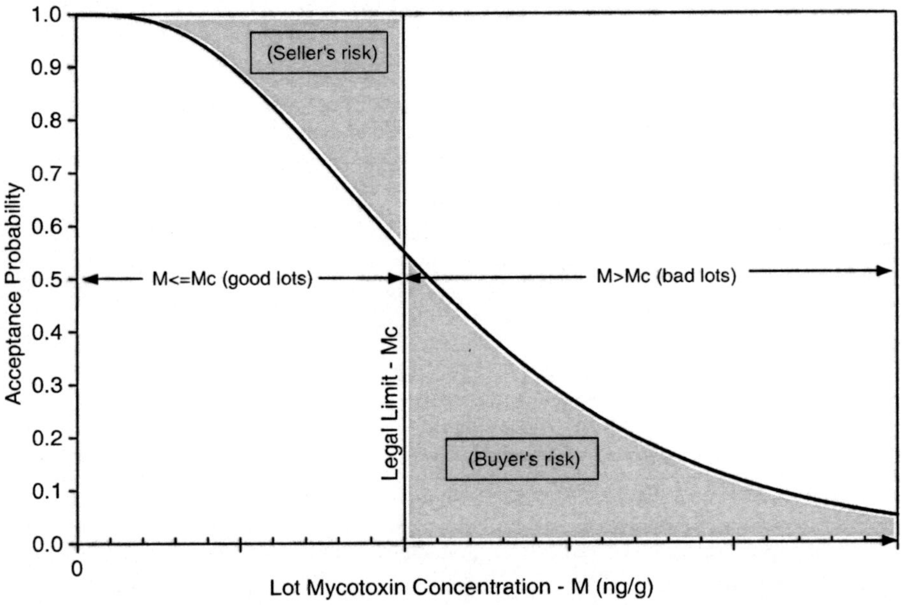

Fig. 10.2 General shape of an operating characteristic (OC) curve. The shape of the OC curve is unique for a mycotoxin test procedure and indicates the magnitude of the buyer's and seller's risks

10.1 Calculation of Acceptance Probability

The acceptance probability P(M) associated with sampling a commodity for a mycotoxin can be computed if the distribution among replicated laboratory sample test results can be described and if the appropriate variance relationships (i.e. Eqs. 8.2, 8.3, and 8.4) are known for a mycotoxin test procedure. Several skewed distributions such as the negative binomial and compound gamma have been shown to adequately describe the observed mycotoxin distribution of sample test results for several commodities and several mycotoxins (Whitaker et al. 1972, 2000; Giesbrecht and Whitaker 1998). An example of how to compute the acceptance probabilities P(M) for aflatoxin sampling plans for shelled corn are described below.

Studies (Johansson et al. 2000b; Giesbrecht and Whitaker 1998) have shown that the compound gamma distribution can accurately describe the distribution of aflatoxin sample test results over a wide range of shelled corn lot concentrations. If a number, n, of replicate laboratory samples of kernels are taken from a lot of shell corn with concentration M, the distribution among sample concentrations can be described by the following function.

$$f_n(x)=e^{-\lambda} \qquad \textbf{if x = 0}$$

$$f_n(x)=\sum_{k=1}^{n}\frac{\lambda^k e^{-\lambda}}{\Gamma(k+1)}\frac{x^{k\alpha-1}e^{-x/\beta}}{\beta^{k\alpha}\Gamma(k\alpha)} \qquad \textbf{if x > 0} \tag{10.1}$$

where x is the total aflatoxin in a laboratory sample of n kernels, Γ is the gamma function, and α, β, and λ are parameters of the compound gamma function. The parameter α controls the shape of the aflatoxin distribution among individual contaminated kernels in the lot, β is a scale parameter, and λ is related to the percent-contaminated kernels in the lot. The parameters α, β, and λ can be calculated from the equations below.

$$\lambda = [(\alpha+1)/\alpha][M^2/VT] \tag{10.2}$$

$$\beta = M/\alpha\lambda \tag{10.3}$$

$$\alpha = 2 \tag{10.4}$$

where VT is the total variance of the aflatoxin test procedure described by Eq. 8.7 and M is the lot aflatoxin concentration.

To calculate the acceptance probabilities or OC curve associated with a sampling plan to detect aflatoxin in shelled corn, one would follow the steps outlined below.

1. **First define the sampling plan or the aflatoxin test procedure: sample size ns,** sample preparation method (type mill and subsample size nss), analytical method (type method and number of aliquots) to be evaluated, number of aliquots quantified na, and the accept/reject limit (X_c). For this example, use the aflatoxin test procedure shown in Table 8.2 where ns = 4.54 kg, Romer mill, nss = 50 g, ELISA, na = 1, and X_c = 20 ng/g.
2. **Calculate the total variance, VT, associated with the aflatoxin test procedure when sampling a lot of shelled corn at a given aflatoxin concentration, M, using** Eq. 9.1. For a lot concentration M = 20 ng/g, VT is

$$\mathbf{VT} = (\mathbf{12.95}/\mathbf{ns})\mathbf{M}^{0.98} + (\mathbf{62.70}/\mathbf{nss})\mathbf{M}^{1.27} + (\mathbf{0.631}/\mathbf{na})\mathbf{M}^{1.293}$$

$$\mathbf{VT} = 53.3 + 56.3 + 27.9 = 137.5$$

3. Calculate the parameters of the compound gamma distribution **α**, **β**, and **λ** from Eqs. 10.2, 10.3, and 10.4.

$$\lambda = [(\alpha + 1)/\alpha][M^2/VT] = [(2+1)/2][20^2/137.5] = 4.36$$

$$\beta = M/\alpha\lambda = 20/(2)(4.36) = 2.29$$

$$\alpha = 2$$

4. Calculate, using the compound gamma distribution in Eq. 10.1, the probability of obtaining a sample test result, X, less than or equal to the accept/reject limit, X_c, of 20 ng/g, $P(X \le X_c|M)$, for M = 20 ng/g, **α** = 2, **β** = 2.29, and **λ** = 4.36.
5. Repeat the process (I through IV) for a range of M values. Choose M values where $P(X \le X_c|M)$ varies from greater than 0.0 to less than 1.0. Record values $P(X \le X_c|M)$ and M values to determine the magnitude of the buyer's and seller's risks.

The probability of accepting and rejecting shelled corn lots over a range of lot concentration for the sampling plan described in part (I) above is shown in Table 10.1. Table 10.1 shows that almost all lots below 5 ng/g are accepted by the sampling plan and almost all lots above 60 ng/g are rejected by the sampling plan. For example, 98% and 2% of the lots at 5 and 60 ng/g are accepted by the sampling

Table 10.1 Probability of accepting and rejecting lots of shelled corn over a range of lot aflatoxin concentrations for a sampling plan that uses 4.54 kg laboratory sample size, Romer mill, 50 g test portion, ELISA method, 1 aliquot, and accept/reject limit of 20 ng/g

Lot concentration M (ng/g)	Probability of accepting lot at M P(X ≤ Xc/M)	Probability of rejecting lot at M 1-P(X ≤ Xc/M)
0	1.000	0.000
5	0.983	0.017
10	0.887	0.113
15	0.726	0.274
20	0.551	0.449
25	0.396	0.604
30	0.274	0.726
35	0.185	0.815
40	0.122	0.878
45	0.079	0.921
50	0.051	0.949
55	0.032	0.968
60	0.020	0.980

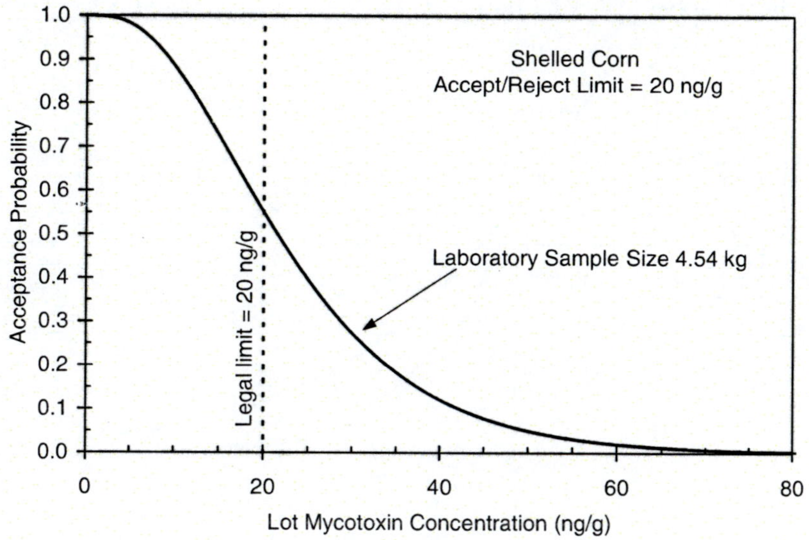

Fig. 10.3 Operating characteristic curve for a sampling plan that uses a 4.54 kg laboratory sample, Romer mill, 50 g test portion, ELISA method, 1 aliquot, and a 20 ng/g accept/reject limit

plan, respectively. As lot concentration M increases, the percentage lots accepted by the sampling plan decreases. The acceptance probabilities in Table 10.1 are plotted in Fig.10.3 and a smooth curve forced through the points.

For a given sampling plan, the OC curve indicates the magnitudes of the buyer's risk and seller's risk. When Mc is defined as the legal limit or the maximum lot concentration acceptable, lots with M > Mc are bad and lots with M <= Mc are good. In Fig. 10.2, the area under the OC curve for M > Mc represents the buyer's risk (bad lots accepted) while the area above the OC curve for M < Mc represents the seller's risk (good lots rejected) for a particular sampling plan design. Using the example in Table 10.1, if lots at 20 ng/g or less are considered good lots and lots greater than 20 ng/g are consider bad lots, then lots rejected below 20 ng/g are considered a measure of the seller's risk (good lots rejected) and the lots accepted above 20 ng/g are considered the buyer's risk (bad lots rejected).

Because the shape of the OC curve is uniquely defined by the laboratory sample size, degree of comminution, test portion size, the type of analytical method, number of aliquots analyzed, and the accept/reject limit, these parameters can be used to reduce the buyer's and seller's risks associated with a sampling plan design.

10.2 Effect of Uncertainty (Sample Size) on Risks

Reducing the uncertainty of the mycotoxin test procedure will reduce the misclassification of lots with any mycotoxin-sampling plan. As demonstrated above, the uncertainty (variability) of a mycotoxin test procedure can be reduced by increasing

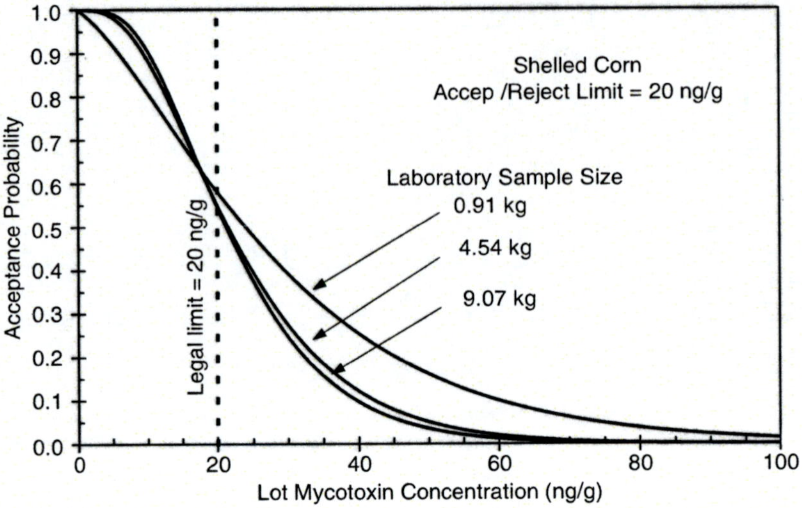

Fig. 10.4 Three operating characteristic curves showing that increasing sample size reduces both buyer's and seller's risks when testing shelled corn for aflatoxin

sample size ns, increasing either the degree of sample comminution, increasing the test portion size nss, and/or increasing the number of aliquots na quantified. The effect of increasing sample size on the shape of the OC curve when testing shelled corn lots for aflatoxin is shown in Fig. 10.4 where the accept/reject limit is equal to the legal limit of 20 ng/g. As sample size increases from 0.91(2 lbs) to 9.07 kg (20 lbs), the slope of the OC curve about legal limit increases forcing the two areas associated with each risk to decrease. As a result, increasing the size sample decreases both the buyer's and seller's risks.

10.3 Effect of Accept/Reject Limit on Risks

The effect of changing the accept/reject limit, relative to the legal limit, on the two risks when testing shelled corn lots for aflatoxin is shown in Fig. 10.5. If the legal limit is assumed to be 20 ng/g, then changing X_c to a value less than 20 ng/g shifts the OC curve to the left. Compared to the sampling plan where X_c = 20 ng/g, the buyer's risk decreases, but the seller's risk increases. If X_c becomes larger than 20, the OC curve shifts to the right. As a result, the seller's risk decreases but the buyer's risk increases. Changing the accept/reject limit relative to the legal limit can reduce only one of the two risks, because reducing one risk will automatically increase the other risk.

Reducing the accept/reject limit below the regulatory limit reduces the buyer's risk, but increases the seller's risk. Often importers when contracting for a shipment specify that the exporter must use an accept/reject limit below the regulatory limit because it reduces the importer's or buyer's risk and forces the exporter or seller to

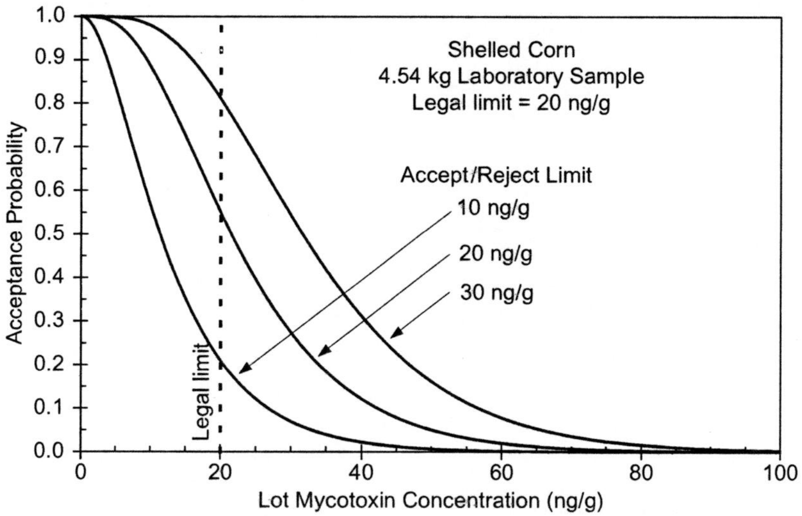

Fig. 10.5 Three operating characteristic curves for three accept/reject limits used to test shelled corn for aflatoxin. If the accept/reject limit (10 ng/g) is less than the legal limit (20 ng/g), the seller's risk increases and the buyer's risk decreases. If the accept/reject (30 ng/g) is greater than the legal limit (20 ng/g), the seller's risk decreases and the buyer's risk increases

take the largest share of the risk. Increasing the accept/reject limit above the regulatory limit increases the buyer's risk, but reduces the seller's risk. It is rare to find an accept/reject limit greater than the regulatory limit, but this situation can be used early in the market system when a handler knows that lot contamination can be reduced by using various sorting methods to remove contaminated product.

Methods have been developed to predict the seller's and buyer's risks, the total number of lots accepted and rejected, the amount of mycotoxin in the accepted and rejected lots, and the costs associated with a mycotoxin inspection program for several commodities (Food and Agriculture Organization 1993; Whitaker and Dickens 1979; Johansson et al. 2000c). These methods have been used by the UDSA/AMS and the peanut industry to design aflatoxin-testing programs for shelled peanuts (Whitaker et al. 1995) and by the FAO (Giesbrecht and Whitaker 1998) to design the aflatoxin-testing plan for raw shelled peanuts destined for further processing.

Chapter 11
Conclusions

Because of the uncertainties (biases and variability) associated with a mycotoxin test procedure, it is impossible to determine with 100% certainty the true concentration of a bulk lot. Even when the sample is correctly selected (no biases), there will be variability associated with the mycotoxin test procedure. The variance associated with a mycotoxin test procedure is the sum of sampling, sample preparation, and analytical variances. For small sample sizes, sampling is usually the largest source of variability. Increasing laboratory sample size, the degree of sample comminution, test portion size, and the number of aliquots quantified can reduce the variability associated with a mycotoxin test procedure. Reducing variability of the mycotoxin test procedure will reduce the number of lots misclassified by the sampling plan. An example of measuring uncertainty and developing a model to predict the performance (OC curves) of sampling plans to detect fumonisin in maize lots (Whitaker et al. 2007) is shown in Appendix A.

T.B. Whitaker et al., *Sampling Procedures to Detect Mycotoxins in Agricultural Commodities*,
DOI 10.1007/978-90-481-9634-0_11,

Chapter 12
References

Association of Official Analytical Chemists (1990) In: Kenneth Helrich (ed) Official methods of analysis of the association of official analytical chemists, 15th edn. Association of Official Analytical Chemists, Inc, Arlington, VA, pp 22201–23301

Bauwin GR, Ryan HL (1982) Sampling inspection and grading of grain. In: Christensen CM (ed) Storage of cereal grains and their products, vol 5. American Association of Cereal Chemistry, St. Paul, Minnesota, p 115

Campbell AD, Whitaker TB, Pohland AE, Dickens JW, Park DL (1986) Sampling, sample preparation, and sampling plans for foodstuffs for mycotoxin analysis. Pure Appl Chem 58:305–314

Cochran WG, Cox GM (1957) Experimental design. Wiley, New York, pp 15–16

Commission regulation (EC) 401/2006 of 23 February 2006 laying down the methods of sampling and analysis for the official control of the levels of mycotoxins in foodstuff. Official Journal of the European Union, L70/12, 9.3.2006

Council for Agricultural Science and Technology (2003) Mycotoxins: risks in plant animal, and human systems. Task Force Report Number 139. Ames, IA, 500014, pp 199

Cucullu AF, Lee LS, Pons WA (1977) Relationship of physical appearance of individual mold damaged cottonseed to aflatoxin content. J Am Oil Chem Soc 54:235A

Cucullu AF, Lee LS, Mayne RY, Goldblatt LA (1986) Determination of aflatoxin in individual peanuts and peanut sections. J Am Oil Chem Soc 43:89

Cullen JM, Newberne PM (1994) Acute hepatotoxicity of aflatoxin. In: Eaton DL, Groopman JD (eds) The toxicology of aflatoxin. Academic, San Diego, CA, pp 3–26

Dickens JW, Satterwhite JB (1969) Subsampling mill for peanut kernels. Food Technol 23:90–92

Dickens JW, Whitaker TB (1982) Sampling and sampling preparation. In: Egan H, Stoloff L, Scott P, Costegnaro M, O'Neill IK, Bartsch H (eds) Environmental carcinogens-selected methods of analysis: some mycotoxins, vol 5. ARC, France, p 17

Dickens JW, Whitaker TB, Monroe RJ, Weaver JN (1979) Accuracy of subsampling mill for granular material. J Am Oil Chem 56:842

Food and Agriculture Organization (1993) Sampling plans for aflatoxin analysis in peanuts and corn. FAO Food and Nutrition Paper 55. FAO, Viale della Terme di Caracalla, 00100, Rome, Italy

Food and Agriculture Organization (2001a) Sampling plan for peanuts. Food and Agriculture Organization, Rome, Italy

Food and Agriculture Organization (2001b) CODEX STAN 209-1999, Rev. 1-2001, Maximum level and sampling plan for total aflatoxins in peanuts intended for further processing. Viale della Terme di Caracalla, 00100, Rome, Italy

Food and Agriculture Organization (2003) Worldwide regulations for mycotoxins in food and feeds in 2003. FAO Food and Nutrition Paper 81. FAO, Viale della Terme di Caracalla, 00100, Rome, Italy, pp165

Food and Agriculture Organization/World Health Organization (2001) Proposed draft revised sampling plan for total aflatoxin in peanuts intended for further processing. Joint FAO/WHO food standards program, CODEX alimentarus commission, 24th session, Geneva, Switzerland,

July 2–7, 2001. FAO/WHO Joint Office, Viale della Terme di Caracalla, 00100, Rome, Italy, pp 276–280

Giesbrecht FG, Whitaker TB (1998) Investigations of the problems of assessing aflatoxin levels in peanuts. Biometrics 54:739–753

Hart LP, Schabenberger O (1998) Variability of vomitoxin in truckloads of wheat in a wheat scab epidemic year. Plant Dis 82:625–630

Hurburgh CR, Bern CJ (1983) Sampling corn and soybeans. 1. Probing method. Trans Am Soc Agric Eng 26:930

Johansson AS, Whitaker TB, Hagler WM Jr, Giesbrecht FG, Young JH, Bowman DT (2000a) Testing shelled corn for aflatoxin, Part I: estimation of variance components. J Assoc Off Anal Chem Int 83:1264–1269

Johansson AS, Whitaker TB, Hagler WM Jr, Giesbrecht FG, Young JH (2000b) Testing shelled corn for aflatoxin, Part II: modeling the distribution of aflatoxin test results. J Assoc Off Anal Chem Int 83:1270–1278

Johansson AS, Whitaker TB, Hagler WM Jr, Giesbrecht FG, Young JH (2000c) Testing shelled corn for aflatoxin, Part III: evaluating the performance of aflatoxin sampling plans. J Assoc Off Anal Chem Int 83:1279–1284

Maestroni BM, Brambilla V, Dabalus-Venida M, Rathor N, Doko B, Sako A, Ambrus A (2005) Estimation of the uncertainty of sampling for the analysis of Fumonisin FB_1 in maize. Proceedings of BCPC Glasgow conference, Glasgow, November 2005

Malone B (2000) Solution fluorometric method for deoxynivalenol in grains. In: Trucksess MW, Pohland AE (eds) Mycotoxin protocols. Humana, Totowa, NJ, pp 97–113

Moss MO (1991) The environmental factors controlling mycotoxin formation. In: Smith JE, Henderson RS (eds) Mycotoxins and animal foods. CRC Press, Boca Raton, FL, pp 37–56

Nesheim S (1979) Methods of aflatoxin analysis. NBS Spec Publ (US) No. 519, pp 355

Nesheim S, Trucksess MW (1986) Thin-layer chromatography/high performance thin-layer chromatography as a tool for mycotoxin determination. In: Cole RJ (ed) Modern methods in the analysis and structural elucidation of mycotoxins. Academic, Orlando, FL, pp 239–264

Parker PE, Bauwin GR, Ryan HL (1982) Sampling, inspection, and grading of grain. In: Christensen CM (ed) Storage of cereal grains and their products. American Association of Cereal Chemists, St. Paul, MN, pp 1–35

Phillips TD, Clement BA, Park DL (1994) Approaches to reduction of aflatoxin in foods and feeds. In: Eaton DL, Groopman JD (eds) The toxicology of aflatoxin. Academic, San Diego, CA, pp 383–406

Remington RD, Schrok MA (1970) Statistics and applications to the biological and health sciences. Prentice-Hall, Englewood Cliffs, NJ, p 105

Schatzki TF (1995a) Distribution of aflatoxin in pistachios. 1. Lot distributions. J Agric Food Chem 43:1561–1565

Schatzki TF (1995b) Distribution of aflatoxin in pistachios. 2. Distribution in freshly harvested pistachios. J Agric Food Chem 43:1566–1569

Shotwell OL, Goulden ML, Hessletine CW (1974) Aflatoxin: distribution in contaminated corn. Cereal Chem 51:492

Shotwell OL, Goulden ML, Botast RJ, Hasseltine CW (1975) Mycotoxins in hot spots in grains. 1. Aflatoxin and zearalenone occurrence in stored corn. Cereal Chem 52:687

Steyn PS, Thiel PS, Trinder DW (1991) Detection and quantification of mycotoxins by chemical analysis. In: Smith JE, Henderson RS (eds) Mycotoxins and animal foods. CRC Press, Boca Raton, FL, pp 165–221

United States Department of Agriculture (1975) Inspectors instructions. Agricultural Marketing Service, Washington, DC, p 22

Whitaker TB, Dickens JW (1979) Evaluation of the peanut administrative committee testing program for aflatoxin in shelled peanuts. Peanut Sci 6:7–9

Whitaker TB, Dowell FE (1995) Sampling methods to measure aflatoxin and grade factors of peanuts. In: Pattee HE, Stalker HT (eds) Advances in peanut science. Am Peanut Res Educ Soc, Stillwater, OK, pp 475–499

Whitaker TB, Dickens JW, Monroe RJ (1972) Comparison of the observed distribution of aflatoxin in shelled peanuts to the negative binomial distribution. J Am Oil Chem Soc 49:590–593

Whitaker TB, Dickens JW, Monroe RJ (1974) Variability of aflatoxin test results. J Am Oil Chem Soc 51:214–218

Whitaker TB, Whitten ME, Monroe RJ (1976) Variability associated with testing cottonseed for aflatoxin. J Am Oil Chem Soc 53:502–505

Whitaker TB, Dickens JW, Monroe RJ (1979) Variability associated with testing corn for aflatoxin. J Am Oil Chem Soc 56:789–794

Whitaker TB, Dowell FE, Hagler WM Jr, Giesbrecht FG, Wu J (1993) Variability associated with sampling, sample preparation, and chemically testing farmers' stock peanuts for aflatoxin. J Assoc Off Anal Chem Int 77:107–116

Whitaker TB, Springer J, Defize PR, deKoe WJ, Coker R (1995) Evaluation of sampling plans used in the United States, United Kingdom, and The Netherlands to test raw shelled peanuts for aflatoxin. J Assoc Off Anal Chem Int 78:1010–1018

Whitaker TB, Horwitz W, Albert R, Nesheim S (1996) Variability associated with methods used to measure aflatoxin in agricultural commodities. J Assoc Off Anal Chem Int 79:476–485

Whitaker TB, Trucksess M, Johansson A, Giesbrecht FG, Hagler WM Jr, Bowman DT (1998) Variability associated with testing shelled corn for Fumonisin. J Assoc Off Anal Chem Int 81:1162–1168

Whitaker TB, Hagler WM Jr, Giesbrecht FG, Johansson AS (2000) Sampling, sample preparation, and analytical variability associated with testing wheat for deoxynivalenol. J Assoc Off Anal Chem Int 83:1285–1292

Whitaker TB, Doko B, Maestroni BM, Slate AB, Ogunbanwo B (2007) Evaluating the performance of sampling plans to detect fumonisins B_1 in maize lots marketed in Nigeria. J Assoc Off Anal Chem Int 90:1050–1059

William PC (1991) Storage of grains and seeds. In: Smith JE, Henderson RS (eds) Mycotoxins and animal foods. CRC Press, Boca Raton, FL, pp 721–746

Chapter 13
Appendix A

Evaluating the Performance of Sampling Plans to Detect Fumonisin B_1 (FB_1) in Maize Lots Marketed in Nigeria

A study was carried out to evaluate the performance of sampling plans to determine fumonisin in maize produced and marketed in Nigeria, Africa (Whitaker et al. 2007). A total of 86 food-grade maize lots intended for human consumption were sampled in 2002 from five regions in Nigeria. From each lot, a 2 kg 'aggregate' sample was taken, comprising 20 laboratory samples of 100 g each. Each laboratory sample was identified by sample number, lot number, and location. Each 100 g laboratory sample was finely ground using a RAS II Romer mill. The comminuted 100 g laboratory sample was thoroughly mixed before removing a 25 g test portion for fumonisin extraction. Fumonisin B_1 was extracted from the 25 g test portion with 50 mL-methanol-water (3 + 1) into a 500 mL Duran screw-cap glass container, using Certomat SII rotary shaker (B. Braun Biotech International), 1 h at 170 rpm, and then filtered through Whatman filter paper number 4. Fumonisin B_1 was analyzed using high performance liquid chromatography (HPLC) with fluorescence detection, using an orthophthalaldehyde (OPA) derivatization method. For each lot, an observed FB_1 distribution was constructed from the 17 laboratory sample test results and a total of 86 observed FB_1 distributions were obtained.

Two theoretical distributions, negative binomial and compound gamma, were chosen as possible models to simulate the observed fumonisin FB_1 distribution among the 17 laboratory sample test results taken from a given lot and the mean and variance for a given lot were used to compute the parameters of the distributions. A goodness of fit (GOF) test was applied to the theoretical distribution, F(f) and the observed distribution $S_n(f)$. The negative binomial model provided suitable fits to more lots than the compound gamma model and therefore the negative binomial was chosen to calculate the operating characteristic (OC) curves, or in other words to calculate the probability of accepting (or rejecting) a lot at a given fumonisin concentration by a specific sampling plan design. All sampling plan designs evaluated in this study used a Romer RAS II mill, a 25 g test portion, and HPLC for quantification of the Fumonisin B_1 concentration.

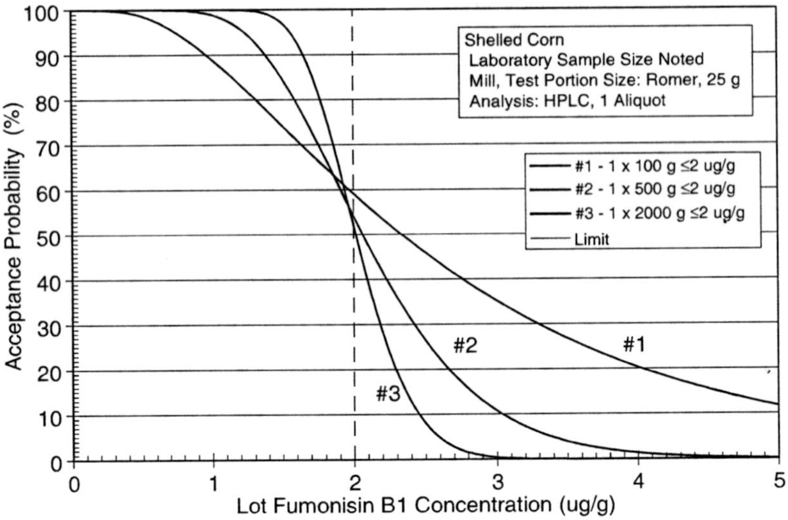

Fig. A.1 Operating characteristic curves for three sampling plan designs that use 100, 500, and 2,000 g samples. The accept/reject limit is 2.0 μ/g

Figure A.1 shows three OC curves reflecting laboratory sample sizes of 100, 500, and 2,000 g. All three sampling plans used an accept/reject limit of 2.0 μg/g. As sample size increases, the OC curves get steeper around the regulatory limit of 2.0 μg/g. The chances of accepting good lots increases and the chances of accepting bad lots decreases; both the buyer's and seller's risks get smaller as sample size increases. Because the sampling step accounts for most of the total variability associated with the fumonisin test procedure, increasing sample size is often the first approach taken to reduce the buyer's and seller's risks.

Changing the accept/reject limit (f_a) relative to the regulatory limit can also be used to reduce either the seller's risk or the buyer's risk, but this approach cannot reduce both risks at the same time. It is important to be able to predict the buyer's and seller's risks associated with a sampling plan used to detect fumonisin in maize so that a sampling plan can be designed to reduce risks associated with misclassifying lots. Once the magnitude of the buyer's and seller's risks are known, sampling plan design parameters, such as laboratory sample size and/or accept/reject limits, can be changed to make the risks more acceptable to the buyer and/or seller of the product being inspected. By changing these sampling plan design parameters, it is possible to adjust the performance of the sampling plan according to risks levels specified by the buyer and seller.

Index